T0259718

# Power Systems

More information about this series at http://www.springer.com/series/4622

Subrata Karmakar · Surajit Chattopadhyay
Madhuchhanda Mitra · Samarjit Sengupta

# Induction Motor Fault Diagnosis

Approach through Current Signature
Analysis

 Springer

Subrata Karmakar
University College of Science, Technology
  and Agriculture
University of Calcutta
Kolkata
India

Surajit Chattopadhyay
Department of Electrical Engineering
Ghani Khan Choudhury Institute of
  Engineering and Technology
Malda
India

Madhuchhanda Mitra
Department of Applied Physics
University of Calcutta
Kolkata, West Bengal
India

Samarjit Sengupta
Department of Applied Physics
University of Calcutta
Kolkata, West Bengal
India

ISSN 1612-1287                   ISSN 1860-4676   (electronic)
Power Systems
ISBN 978-981-10-9213-8           ISBN 978-981-10-0624-1   (eBook)
DOI 10.1007/978-981-10-0624-1

Printed on acid-free paper

This Springer imprint is published by Springer Nature
The registered company is Springer Science+Business Media Singapore Pte Ltd.

To
The Spouses of the Authors

# Preface

Induction motor applications in industry are becoming more and more complex. The dependence of drives and automation on induction motor has made the control and the fault detection of the motor sufficiently intricate. With the increase of load variation and use of modern power electronic devices, nowadays the complexities in construction and operation are made more efficient, user-friendly, and reliable also. But the usage of these devices has pushed motors into a fault-prone environment. Since the use of sophisticated electronic gadgets has increased in every sphere of life, for their good longevity, requirement of early motor fault diagnosis has become a predominant criterion to the consumers in the competitive market of drives and automation. Therefore, early fault diagnosis has become the concern of utilities, end users, operating engineers, as well as manufacturers. This book is intended for graduates, postgraduates, and researchers, as well as for professionals in the related fields.

This book has evolved from the researches carried out by the authors and the contents of the courses are given by the authors at University of Calcutta, Department of Applied Physics, India in the bachelor's and master's courses in electrical engineering. A large number of references are given in the book, most of which are journal and conference papers and national and international standards.

The content of the book focuses, on the one hand, on different motor faults, their sources, and effects, and on the other hand, different analytical methods for diagnosis of various motor faults. Advantages and limitations of different methods are discussed along with simulated and laboratory experiment results. At the end, a chapter has been added to focus on the general discussion of the induction motor fault diagnosis and research scope.

The key features of the book can be highlighted as follows:

- This book has approached the subject matter in a lucid language. Fault diagnosis techniques have their analytical background supplemented by simulated and experimental results.

- This book has mainly handled fault diagnosis methods using both steady-state motor current and starting current transients, which are absent in many other similar books.
- In general, the book has dealt with common motor faults which are required for students, researchers, and practicing engineers.
- The content level of the book is designed in such a way that the general description of different induction motor faults are made first, followed by some existing and new fault diagnosis techniques. This content should attract the students, researchers, and practicing engineers.
- The predominant features of the book are

  – Lucid but concise description of the subject (which may be available in other books).
  – Chapter outcome in each chapter is clearly mentioned.
  – Abstract in each chapter is provided which will attract the researchers.
  – Detailed new measurement techniques (which are not available in other books).
  – Simulation results of new techniques and comparison of the same with experimentation will give fresh researchers a guideline as to how the can be done (which are not available in other books).

The authors wish to thank members of the Springer publisher for their support to publish the book.

Authors owe a particular debt of gratitude to the teachers of Department of Applied Physics for their constant support. Last but not least, the authors are indebted to their parents, spouses, and children, without whose constant endurance, this book would not have seen the light.

November 2015

Subrata Karmakar
Surajit Chattopadhyay
Madhuchhanda Mitra
Samarjit Sengupta

# Contents

# About the Authors

**Subrata Karmakar** born in the year 1966 in West Bengal, India completed his graduation with Honours in Physics from the Ramakrishna Mission Residential College, Narendrapur in West Bengal under University of Calcutta, Kolkata, India. Subsequently, he obtained B.Tech and M.Tech degrees in Electrical Engineering and Ph.D. (technology) from the same university. His employment experience includes both private and government sectors. In private sector he worked as Sr. Electrical Engineer (Design). He was in National Institute of Technology, Sikkim, India as Faculty In-charge and Assistant Professor (ad hoc). Now he is associated with the University of Calcutta as Assistant Secretary, University College of Science, Technology and Agriculture. His special field of interest includes motor fault analysis. He has published 15 papers in journals and conferences.

**Surajit Chattopadhyay** has obtained B.Sc. Degree in Physics Honours from Ramakrishna Mission Vidyamandir (Belur Math), University of Technology in 1998, and then B.Tech., M.Tech., and Ph.D. (Technology) degrees in Electrical Engineering from the Department of Applied Physics of University of Calcutta in 2001, 2003, and 2010, respectively. He has obtained CEng from Engineering Council, UK in 2013. He has authored/coauthored around 97 papers published in international and national journals and conferences and three books. Seven papers have been selected as "Best Paper" at international level. He has visited many countries such as Lyon (France), Kuala Lumpur (Malaysia), Dhaka (Bangladesh), London and Stevenage (UK), and Negembo (Sri Lanka) for technical interaction and presented his work in different international forums. Currently, he is Dean (Student Welfare) and Associate Professor of Electrical Engineering in Ghani Khan Chaoudhury Institute of Engineering and Technology (under Ministry of HRD, Government of India). He is Hon. Secretary of the Institution of Engineering and Technology (UK), Kolkata Network since 2013. His field of interest includes electric power quality, fault diagnosis, power system protection, signal analysis, robotics application, and UAV.

**Dr. Madhuchhanda Mitra** was born in Kolkata, India in 1961. She received her B.Tech., M.Tech., and Ph.D. (Tech) degrees in 1987, 1989, and 1998, respectively, from University of Calcutta, Kolkata, India. At present, she is Professor in the Department of Applied Physics, University College of Technology, University of Calcutta, India, where she actively engages in both teaching and research. She is the coauthor of 150 research papers pertaining to the problems of fault analysis, biomedical signal processing, data acquisition, and processing and materials science. Her total citation is 1088, h-index is 15, and i-10 index is 24. She is a recipient of "Griffith Memorial Award" of the University of Calcutta.

**Samarjit Sengupta** holds B.Sc, B.Tech, M.Tech, and Ph.D. degrees from the University of Calcutta, Kolkata, India. He is currently a Professor of Electrical Engineering in the Department of Applied Physics at the University of Calcutta. He has published 130 journal papers and eight books on various topics of electrical engineering. His main research interests include power quality instrumentation, power system stability, and security and power system protection. He is a fellow of IET and IETE, as well as a senior member of IEEE. He is former Chairman of IET (UK) Kolkata Network.

# List of Figures

# List of Tables

# Chapter 1
# Introduction

**Abstract** The chapter deals with general introduction of the book. First, it mentions the importance of fault diagnosis. Then the role of induction motor and different aspects of induction motor fault analysis have been discussed. Main objective of the book has been mentioned. At the end, books at a glance has been presented with the help of short chapter highlights.

**Keywords** Induction motor · Fault diagnosis · Objective

**Chapter Outcome**

After completion of the chapter, readers will be able to gather knowledge and information regarding the following areas:

- Importance of motor fault diagnosis
- Induction motor faults
- Highlights of different approaches for motor fault diagnosis
- Different analytical tools used in motor fault diagnosis
- Object of the book
- Chapter at a glance.

## 1.1 Importance of Fault Diagnosis

In this era of industrial automation, remarkable advancement has taken place in processor-based signal acquisition and analysis and also in knowledge-based development. On the basis of these developments, conditioning monitoring of electrical machines is being done. The area of system maintenance cannot realize its full potential if it is only limited to preventive approaches. Rather, the early diagnosis of a developing fault is necessary to allow maintenance personnel to schedule repairs—prior to an actual failure. During the last decades, there has been much interest in early fault detection and diagnosis technique for use in condition-based maintenance (CBM). In contrast to preventive maintenance, in CBM, one does not require to schedule maintenance or machine replacement based on previous records or statistical estimates

© Springer Science+Business Media Singapore 2016
S. Karmakar et al., *Induction Motor Fault Diagnosis*,
Power Systems, DOI 10.1007/978-981-10-0624-1_1

of machine failure. Rather, one relies on information provided by condition monitoring system, which assesses the condition of the system. This allows better utilization of components and equipment, leading to considerable reduction of downtime and maintenance cost. For CBM one has to know the changes which may occur in the induction motor parameters and for this, knowledge on fault diagnosis of induction motor is very important, but before that one has to know in detail the construction of an induction motor.

## 1.2  Induction Motor

The main parts of an induction motor are stator, rotor, and ball bearing. Electrical supply is provided to the stator. This stator consists of (i) the outer cylindrical frame of the motor made either of welded sheet steel, cast iron, or cast aluminum alloy; (ii) the magnetic path, called core which comprises a set of slotted laminations which is made up of magnetic substance and is pressed into the cylindrical space inside the outer frame. The core is laminated to reduce eddy currents and by this to reduce losses and heating; and (iii) a set of insulated electrical windings, which are placed inside the slots of the laminated core. The cross-sectional area of these windings depends on the power rating of the motor. For a three-phase motor there are normally three sets of stator windings, one for each phase. Generally, stator contains feet or a flange for mounting of the motor.

The second part called rotor is the rotating part of the motor and is placed inside the stator coaxially. Mechanical power is obtained from this rotor. It consists of (i) a set of slotted laminations made up of magnetic substance and are pressed together in the form of a cylindrical magnetic path and (ii) the electrical circuit. The electrical circuit of the rotor is either wound-rotor type, which comprises three sets of insulated windings with connections brought out to three slip rings mounted on the shaft of the motor. The external connections to the rotating part are made via brushes onto the slip rings, for which this type of motor is also called a slip ring motor. Rotor may be squirrel-cage type also, which comprises a set of copper or aluminum bars installed into the slots, which are connected to an end ring at each end of the rotor. The construction of these rotor windings resembles a 'squirrel cage.' Aluminum rotor bars are usually die cast into the rotor slots, which results in a very rugged construction. Even though the aluminum rotor bars are in direct contact with the steel laminations, practically all the rotor current flows through the aluminum bars and not in the laminations.

Next main part of an induction motor is the ball bearing. There are two set of bearings to support the rotating shaft, one set at driving end and the other at non-driving end. Ball/roller bearings are used for properly placing the rotor inside the stator bore and for minimizing the friction to run the rotor smoothly.

Besides the above three, the other parts that a motor consists are steel shaft for transmitting the torque to the load, cooling fan located at the non-driving end to provide forced cooling for the stator and rotor, and terminal box on top or either side to receive the electrical power supply connections.

Above discussion reveals that construction of induction motor especially squirrel-cage motor is very simple and it is almost unbreakable. Its cost is low; it is highly reliable and rugged. It has very high efficiency, but main drawback is that induction motor is susceptible to different types of faults. Present-day situations demand rotor of electrical machine to be lighter in weight and faster in speed but tighter in tolerances and for these reasons, rotating machinery is becoming increasingly complex. With this increase in complexity, it is important to eliminate as many sources of faults as possible.

Operators of electrical motors are under continual pressure to reduce maintenance cost and to prevent unscheduled downtime of motors. To reduce the downtime and also for reliable operation, early detection of motor faults is highly demanding. For this, fault diagnosis of induction motor has become a burning topic to electrical technologists for last two decades. Normally, current signature or/and vibration signature are analyzed and useful information are provided to the operator, maintainer, and designer regarding the health condition of the motor.

Although induction motor is reliable, rugged, and almost unbreakable, yet they are susceptible to different types of faults. The effects of such faults in induction motors include unbalanced stator currents and voltages, oscillations in torque, reduction in efficiency and torque, overheating, and excessive vibration. Moreover, these faults can increase the magnitude of certain harmonic components of currents and voltages. Induction motor performance may be affected by the following type of faults:

(a) Electrical-related faults
(b) Mechanical-related faults
(c) Environmental-related faults.

Main faults in induction motors may be listed as follows—(i) broken bar fault, (ii) bearing fault, (iii) rotor mass unbalance fault, (iv) bowed rotor fault, (v) stator winding fault, (vi) rotor winding fault, (vii) single phasing fault, and (viii) crawling.

Electrical Power Research Institute (EPRI) works on to find out the percentage occurrence of different types of motor faults due to bearing fault, stator winding fault, rotor-related fault, and all other remaining faults.

## 1.3  Induction Motor Fault Analysis

Different research works are going on to study various types of motor faults at various research places. Different fault detection techniques have been introduced by the researchers. Some methods are based on analysis of nonelectrical parameters. These include vibration signal analysis to detect electrical fault, analysis of air gap flux to detect motor eccentricity, and acoustic diagnosis technique to detect machine insulation fault. Vibration signal has been analyzed using neural network for electrical fault detection, and thermal analysis of induction motor has been done to study fault of an induction motor.

## 1.4   Current Signature Analysis

Motor fault diagnosis may be done by motor current signature analysis (MCSA) or by motor vibration analysis (MVA) or by analysis of both MCSA and MVA. A number of MCSA-based approaches have already been proposed to detect different types of internal mechanical faults like broken rotor bar fault, bearing fault, and mechanical unbalanced rotor fault. Detection of stator voltage unbalances and single phasing effects using advanced signal processing techniques has been described in some other MCSA-based approaches. Online stator current monitoring system has been used for motor fault detection.

Most of the analysis used for fault diagnosis which was started about three decades ago had been performed using fast Fourier transform (FFT)-based tools on the motor current or vibration signature. However, FFT has some limitations like masking of characteristic frequencies (generally small frequency) by supply frequency, inappropriateness for transient signal, etc. To overcome these limitations different new techniques are being used presently.

Some of the presently used signal processing techniques are short-time Fourier transform (STFT), wavelet transform (WT)—both discrete wavelet transform (DWT) and continuous wavelet transform (CWT)—wavelet packet decomposition (WPD), Wigner Ville distribution (WVD), power spectral density (PSD), Park transform, SVM, Prony analysis, fractal, and fuzzy logic. Most of these techniques use nonstationary signal. In motor, starting current as well as steady-state current has been used to detect fault due to broken rotor bar. Motor internal faults have been characterized using finite element and DWT. Motor signature analysis has been performed using PSD and WPD. Also, Park transform and neural network have been used to analyze stator current signature.

FFT gives different frequency components present in the signal that is being analyzed. In case of motor data analysis, FFT transforms data from time domain to frequency domain which needs accurate slip estimation for frequency component localization in any spectrum. Also it is seen that in case of different faults of motor, frequencies generated are very near to the fundamental frequency with small amplitude. Thus characteristic frequencies for the faulted motor being very close to the fundamental component, and their amplitude being small in comparison to the fundamental, detection of fault, and determination of fault severity under light load is not possible specially for small motors. Also due to variation of motor load, inertia, torque, supply voltage, or speed oscillation of motor, some small harmonics may generate which are similar to the characteristic frequencies for the faulted motor. For this, diagnosis of motor fault by observation of motor current frequencies will be confusing and wrong. A significant drawback of FFT analysis is that it cannot differentiate the harmonics generated due to motor fault from those appearing due to either load variation or voltage fluctuation or speed oscillation. This problem was sorted out by the use of STFT method.

STFT is capable to analyze transient signal. STFT uses constant-sized window to analyze all frequencies—this is the limitation of this method. This limited window may find it difficult to match the frequency content of the signal which is generally not known prior to the analysis. To overcome, this limited sized window is required to be replaced by a variable sized window—WT does it suitably.

WT is an advanced powerful mathematical tool. It is suitable for analyzing transient signal. WT decomposes a signal in both time and frequency in terms of a wavelet, called mother wavelet. WT has also some drawbacks, e.g., selection of mother wavelet is quite arbitrary—this may introduce error in the detection parameters. For lower order wavelet overlap between bands and frequency response will be very poor. Some parts of fundamental frequency leaked into adjacent frequency bands to mask the lower side harmonics. Further in some cases, the edge distortion from the transform makes the detection of the lower frequency band (i.e., below supply frequency) difficult, especially when the starting transient is very fast. To overcome these constraints a new methodology called Hilbert transform (HT) has been proposed.

This HT method has overcome the problem that may arise in WT due to improper selection of mother wavelet using the envelope analysis of the signal. Even the motor current signal at steady state has been analyzed by this HT method.

## 1.5 Objective of the Book

Objective of this book is to discuss about fault diagnosis of a three-phase induction motor by the analysis of its current signature. Different types of faults may occur in an induction motor. In this book five type of faults, namely (i) broken rotor bar, (ii) rotor mass unbalance fault, (iii) stator winding fault, (iv) single phasing, and (v) crawling of an induction motor, will be studied. For diagnosis purpose both steady-state current and transient current of motor will be considered as data to analyze. For analysis, signal processing tools, namely FFT, DWT, CWT, HT, Park's vector matrix, feature pattern extraction method (FPEM), and CMS rule set, will be used. Fault diagnosis will also be done by Concordia and radar analysis-based techniques.

## 1.6 Books at a Glance

The book covers totally nine chapters. This chapter is the introduction of the book.

Chapter 2 deals with induction motor and faults. Construction and operation of an induction motor is discussed. Then, different faults of an induction motor are mentioned along with their causes and effects. Mainly, following faults will be discussed in this chapter: broken rotor bar fault, air gap eccentricity including rotor mass unbalance fault, stator fault, single phasing, and crawling.

In Chap. 3, different existing techniques for fault analysis, namely thermal analysis, chemical analysis, acoustic analysis, torque analysis, induced voltage analysis, partial discharge analysis, vibration analysis, and current analysis, are reviewed. Then, different signal processing tools for fault analysis, namely FFT, STFT, WT, HT, and Radar analysis of stator current Concordia, are reviewed. At the end of this chapter, research trend of different analytical methods that are used in fault analysis is described.

In Chap. 4, rotor broken bar has been studied. The chapter includes diagnosis through radar analysis of stator current Concordia and envelope analysis of stator current using Hilbert and WT.

In Chap. 5, rotor mass unbalance has been assessed. Following methods are discussed in the chapter: (i) analysis of starting current using WT, (ii) analysis using WT of the motor starting current at no load, (iii) analysis of vibration and motor current signatures at steady state, and (iv) radar analysis of stator current Concordia at starting.

In Chap. 6, assessment of stator winding fault has been discussed. Different approaches useful for diagnosis of stator winding fault are discussed followed by conclusion.

In Chap. 7, single phasing fault has been discussed. Single phasing is assessed by analysis of phase angle shift, negative sequence components, Concordia, etc.

In Chap. 8, crawling of induction motor has been discussed. Diagnosis of crawling by feature pattern extraction and current Concordia of stator current are discussed.

In Chap. 9, general discussion on the previous chapters is done and then research scope is highlighted.

# Chapter 2
# Induction Motor and Faults

**Abstract** The chapter deals with general description of an induction motor followed by different faults. First, construction of induction motor has been discussed. Then a review of induction motor fault has been presented. Faults like rotor broken bar, mass unbalance, stator faults, single phasing, crawling, bearing faults, etc. are discussed along with causes and effects.

**Keywords** Bearing fault · Broken rotor bar · Construction · Crawling · Induction motor · Mass unbalance · Single phasing · Stator fault

**Chapter Outcome**

Aftercompletion of the chapter, readers will be able to gather knowledge and information regarding the following areas:

- Construction of induction motor
- Different classes of induction motor
- Different motor faults
- Statistics on motor fault
- Broken rotor bar
- Rotor mass unbalance
- Stator winding fault
- Single phasing
- Crawling
- Bearing fault
- Over/under voltage, overload.

## 2.1  Introduction

An induction motor comprises a magnetic circuit interlinking two electric circuits which are placed on the two main parts of the machine: (i) the stationary part called the stator and (ii) the rotating part called the rotor. Power is transferred from one

© Springer Science+Business Media Singapore 2016
S. Karmakar et al., *Induction Motor Fault Diagnosis*,
Power Systems, DOI 10.1007/978-981-10-0624-1_2

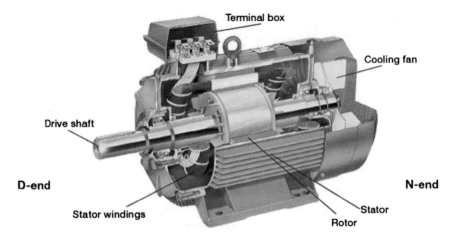

**Fig. 2.1** An induction motor (dissected)

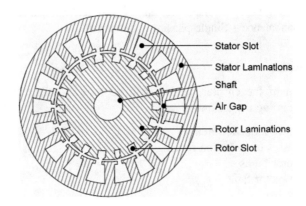

**Fig. 2.2** Magnetic circuit of stator and rotor of an induction motor

part to the other by electromagnetic induction. For this induction machine is
referred as an electromechanical energy conversion device which converts electrical
energy into mechanical energy [1]. Rotor is supported on bearings at each end.
Generally, both the stator and rotor consist of two circuits: (a) an electric circuit to
carry current and normally made of insulated copper or insulated aluminum and
(b) a magnetic circuit, shown in Fig. 2.2, to carry the magnetic flux made of
laminated magnetic material normally steel (Fig. 2.1).

## 2.2   Construction

### (a)  Stator

The stator, shown in Fig. 2.3, is the outer stationary part of the motor. It consists of (i) the outer cylindrical frame, (ii) the magnetic path, and (iii) a set of insulated electrical windings.

(i) **The outer cylindrical frame**: It is made either of cast iron or cast aluminum alloy or welded fabricated sheet steel. This includes normally feet for foot mounting of the motor or a flange for any other types of mounting of the motor.

(ii) **The magnetic path**: It comprises a set of slotted high-grade alloy steel laminations supported into the outer cylindrical stator frame. The magnetic path is laminated to reduce eddy current losses and heating.

(iii) **A set of insulated electrical windings**: For a 3-phase motor, the stator circuit has three sets of coils, one for each phase, which is separated by 120° and is excited by a three-phase supply. These coils are placed inside the slots of the laminated magnetic path.

### (b)  Rotor

It is the rotating part of the motor. It is placed inside the stator bore and rotates coaxially with the stator. Like the stator, rotor is also made of a set of slotted thin sheets, called laminations, of electromagnetic substance (special core steel) pressed together in the form of a cylinder. Thin sheets are insulated from each other by means of paper, varnish [2]. Slots consist of the electrical circuit and the cylindrical electromagnetic substance acts as magnetic path. Rotor winding of an induction motor may be of two types: (a) squirrel-cage type and (b) wound type. Depending on the rotor winding induction motors are classified into two groups [1–3]: (i) squirrel-cage type induction motor and (ii) wound-rotor type induction motor.

(i) **Squirrel-cage type induction motor**: Here rotor comprises a set of bars made of either copper or aluminum or alloy as rotor conductors which are embedded

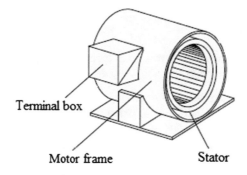

**Fig. 2.3** Stator of an induction motor

**Fig. 2.4**  Squirrel-cage rotor

**Fig. 2.5**  Slip ring rotor

in rotor slots. This gives a very rugged construction of the rotor. Rotor bars are connected on both ends to an end ring to make a close path. Figure 2.4 shows a squirrel-cage type rotor.

(ii) **Wound-rotor type induction motor**: In this case rotor conductors are insulated windings which are not shorted by end rings but the terminals of windings are brought out to connect them to three numbers of insulated slip rings which are mounted on the shaft, as shown in Fig. 2.5. External electrical connections to the rotor are made through brushes placed on the slip rings. For the presence of these slip rings this type of motor is also called slip ring induction motor.

Besides the above two main parts, an induction motor consists some other parts which are named as follows:

(i) **End flanges**: There are two end flanges which are used to support the two bearings on both the ends of the motor.

(ii) **Bearings**: There are two set of bearings which are placed at both the ends of the rotor and are used to support the rotating shaft.

(iii) **Shaft**: It is made of steel and is used to transmit generated torque to the load.

(iv) **Cooling fan**: It is normally located at the opposite end of the load side, called non-driving end of the motor, for forced cooling of the both stator and rotor.

(v) **Terminal box**: It is on top or either side of the outer cylindrical frame of stator to receive the external electrical connections.

## 2.3 Operation

When the stator winding of an induction motor is connected to a three-phase supply, a uniform rotating magnetic field is produced therein [3], which induces e.m.f. in the rotor which is free to rotate coaxially with the stator core with the help of ball bearings. Rotor being short circuited, either through the end rings or an external resistance, currents are produced due to this induced e.m.f. This current interacts with the rotating magnetic field to develop a torque on the rotor in the direction of the rotating magnetic field. As the rotor is free to rotate, the torque will cause it to move round in the direction of the stator field. This makes a three-phase induction motor as self-starting.

In transforming this electrical energy into mechanical energy, in an induction motor some losses occur which are as follows:

- Friction and windage losses, 5–15 %
- Iron or core losses, 15–25 %
- Stator losses, 25–40 %
- Rotor losses, 15–25 %
- Stray load losses, 10–20 %.

Full-load motor efficiency varies from about 85 to 97 %.

Induction motors are simpler, cheaper, and efficient. Among them squirrel-cage induction motor is more rugged and work more efficiently compared to wound-rotor induction motor. If supply voltage and frequency are constant, then a squirrel-cage induction motor runs at a constant speed which makes it suitable for use in constant speed drive [1, 2]. Several standard designs of squirrel-cage induction motors are available in the market to fulfill the requirements of different starting and running conditions of various industrial applications. These are classified [4] as class A, class B, class C, and class D. In Table 2.1, a comparison of different classes of squirrel-cage induction motors is presented.

**Table 2.1** Various classes of squirrel-cage induction motor

|  | Class A | Class B | Class C | Class D |
|---|---|---|---|---|
| Properties | Normal starting torque, high starting current and low operating slip | Normal starting torque, low starting current and low operating slip | High starting torque and low starting current | High starting torque, low starting current and high operating slip |
| Uses | Fan, pump load etc. where torque is low at start | For constant speed drive such as pump, blower | Compressor, conveyors, crashers etc. | For driving intermittent load, e.g. punch press etc. |

**Table 2.2**  Statistics on motor faults/failures [8]

| Type of faults | Number of faults/failures | | | | |
|---|---|---|---|---|---|
| | Induction motor | Synchronous motor | Wound-rotor motor | DC motor | All total motors |
| Bearing | 152 | 2 | 10 | 2 | 166 |
| Winding | 75 | 16 | 6 | – | 97 |
| Rotor | 8 | 1 | 4 | – | 13 |
| Shaft | 19 | – | – | – | 19 |
| Brushes or slip ring | – | 6 | 8 | 2 | 16 |
| External device | 10 | 7 | 1 | – | 18 |
| Others | 40 | 9 | – | 2 | 51 |

## 2.4  Faults: Causes and Effects

Induction motors are rugged, low cost, low maintenance, reasonably small sized, reasonably high efficient, and operating with an easily available power supply. They are reliable in operations but are subject to different types of undesirable faults. From the study of construction and operation of an induction motor, it reveals that the most vulnerable parts for fault in the induction motor are bearing, stator winding, rotor bar, and shaft. Besides due to non-uniformity of the air gap between stator-inner surface and rotor-outer surface motor, faults also occur. Different studies have been performed so far to study reliability of motors, their performance, and faults occurred [5, 6]. The statistical studies of IEEE and EPRI (Electric Power Research Institute) for motor faults are cited in [7, 8]. Part of these studies was to specify the percentage of different faults with respect to the total number of faults. The study of IEEE was carried out on various motors in industrial applications. As per the IEEE Standard 493-1997 the most common faults and their statistical occurrences are shown in Table 2.2. Under EPRI sponsorship, a study was conducted by General Electric Company on the basis of the report of the motor manufacturer. As per their report the main motor faults are presented in Table 2.3 [7, 9].

Faults in induction motors can be categorized as follows:

(a) **Electrical-related faults**: Faults under this classification are unbalance supply voltage or current, single phasing, under or over voltage of current, reverse phase sequence, earth fault, overload, inter-turn short-circuit fault, and crawling.

(b) **Mechanical-related faults**: Faults under this classification are broken rotor bar, mass unbalance, air gap eccentricity, bearing damage, rotor winding failure, and stator winding failure.

(c) **Environmental-related faults**: Ambient temperature as well as external moisture will affect the performance of induction motor. Vibrations of machine, due to any reason such as installation defect, foundation defect, etc., also will affect the performance.

**Table 2.3** Fault occurrence possibility on induction motor [7, 9]

| Studied by | Bearing fault (%) | Stator fault (%) | Rotor fault (%) | Others (%) |
|---|---|---|---|---|
| IEEE | 42 | 28 | 8 | 22 |
| EPRI | 41 | 36 | 9 | 14 |

Faults shown in Table 2.3 are in broad sense; stator fault may be of different kinds, and different types of faults may occur in rotor itself. For identification, faults in induction motors may be listed as follows—(i) broken bar fault, (ii) rotor mass unbalance fault, (iii) bowed rotor fault, (iv) bearing fault, (v) stator winding fault, (vi) single phasing fault, etc. Besides, the phenomenon called crawling when motor does not accelerate up to its rated speed but runs at nearly one-seventh of its synchronous speed is also considered as a fault of an induction motor. Faults listed (i)–(iii) are in general stated as rotor fault which contributes about 8–9 % of the total motor fault. In this work, broken bar fault, rotor mass unbalance fault, stator winding fault, single phasing fault, and crawling are considered.

In an induction motor multiple faults may occur simultaneously and in that case determination of the initial problem is quite difficult [10]. Effects of such faults in induction motor result in unbalanced stator currents and voltages, oscillations in torque, reduction in efficiency and torque, overheating, and excessive vibration [11]. Moreover, these motor faults can increase the magnitude of certain harmonic components of currents and voltages. Induction motor performance may be affected by any of the faults. In the next few paragraphs, causes and effects of different faults in induction motors are discussed.

## 2.5 Broken Rotor Bar Fault

### 2.5.1 General Description of Broken Rotor Bar

The squirrel cage of an induction motor consists of rotor bars and end rings. If one or more of the bars is partially cracked or completely broken, then the motor is said to have broken bar fault. Figure 2.6 shows rotor and parts of broken rotor bar.

### 2.5.2 Causes of Broken Rotor Bar

There are a number of reasons for which rotor faults may occur in an induction motor [12]. It has been observed that in squirrel-cage induction motor rotor asymmetry occurs mainly due to manufacturing defect, such as during the brazing process nonuniform metallurgical stresses may occur in cage assembly which led to failure during rotation of the rotor. Also heavy end rings of rotor result in large centrifugal forces which may cause extra stresses on the rotor bars. Because of any

**Fig. 2.6** Photograph of rotor and parts of broken rotor bar [16]

of the reasons rotor bar may get damage which results in asymmetrical distribution of rotor currents. Also, for such asymmetry or for long run of the motor if any of the rotor bar gets cracked overheating will occur in the cracked position which may lead to breaking of the bar. Now if one of the bars breaks, the side bars will carry higher currents for which larger thermal and mechanical stresses may happen on these side bars. If the rotor continues to rotate in this condition, the side bars may also get cracked [13]—thus damage may spread, leading to fracture of multiple bars of the rotor. This cracking may occur at various locations of the rotor, such as in bars, in end rings, or at the joints of bars and end rings. Possibility is more at the joints of bars and end rings. Moreover, possibilities of crack increase if motor start-up time is long and also if motor is subject to frequent starts and stops [14].

The main causes of rotor broken bar of an induction motor can be mentioned, pointwise, as follows:

- manufacturing defects
- thermal stresses
- mechanical stress caused by bearing faults
- frequent starts of the motor at rated voltage
- due to fatigue of metal of the rotor bar.

### 2.5.3  Effect of Broken Rotor Bar

Cracked or broken bar fault produces a series of sideband frequencies [15, 16], in the stator current signature given by

$$f_{\text{brb}} = f(1 \pm 2ks) \tag{2.1}$$

where $f$ is the supply frequency, $s$ is the slip, and $k$ is an integer.

This has been demonstrated as ripple effect in [16], which explains that the lower side band at $f(1–2s)$ is the strongest which will cause ripples of torque and speed at a frequency of $2sf$ and this in turn will induce an upper side band at $f(1 + 2s)$ and this effect will continue to create the above series of sidebands, i.e., $f(1 \pm 2ks)$. Magnitude of this lower sideband $f(1 - 2s)$ over the fundamental can be used as an indicator of rotor broken bar fault [17].

## 2.6 Rotor Mass Unbalance

From the knowledge of construction of motor it is known that rotor is placed inside the stator bore and it rotates coaxially with the stator. In a healthy motor, rotor is centrally aligned with the stator and the axis of rotation of the rotor is the same as the geometrical axis of the stator. This results in identical air gap between the outer surface of the rotor and the inner surface of the stator. However, if the rotor is not centrally aligned or its axis of rotation is not the same as the geometrical axis of the stator, then the air gap will not be identical and the situation is referred as air-gap eccentricity. In fact air-gap eccentricity is common to rotor fault in an induction motor. Air-gap eccentricity may occur due to any of the rotor faults like rotor mass unbalance fault, bowed rotor fault, etc. Due to this air-gap eccentricity fault, in an induction motor electromagnetic pull will be unbalanced. The rotor side where the air gap is minimum will experience greater pull and the opposite side will experience lower pull and as a result rotor will tend to move in the greater pull direction across that gap [18]. The chance of rotor pullover is normally greatest during the starting period when motor current is also the greatest. In severe case rotor may rub the stator which may result in damage to the rotor and/or stator [19]. Air-gap eccentricity can also cause noise and/or vibration.

### 2.6.1 General Description of Rotor Mass Unbalance

This rotor mass unbalance occur mainly due to manufacturing defect, if not may occur even after an extended period of operation, for nonsymmetrical addition or subtraction of mass around the center of rotation of rotor or due to internal misalignment or shaft bending due to which the center of gravity of the rotor does not coincide with the center of rotation. In severe case of rotor eccentricity, due to unbalanced electromagnetic pull if rotor rubs the stator then a small part of material of rotor body may wear out which is being described here as subtraction of mass, resulting in rotor mass unbalance fault. Figure 2.7 shows rotor mass unbalance fault.

**Fig. 2.7** Rotor with mass unbalance fault

## 2.6.2 Classification of Mass Unbalance

There are three types of mass unbalanced rotor:

(a) Static mass unbalanced rotor
(b) Couple unbalance rotor
(c) Dynamic unbalance rotor.

### 2.6.2.1  Static Mass Unbalanced Rotor

For this fault shaft rotational axis and weight distribution axis of rotor are parallel but offset, as shown in Fig. 2.8. Without special equipment this type of eccentricity is difficult to detect [20].

### 2.6.2.2  Couple Unbalance Rotor

It is shown in Fig. 2.9. If this fault occurs then the shaft rotational axis and weight distribution axis of rotor intersect at the center of the rotor.

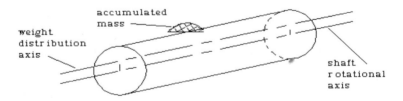

**Fig. 2.8** Static mass unbalanced rotor

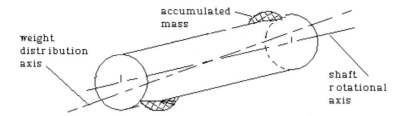

**Fig. 2.9**  Couple unbalanced rotor

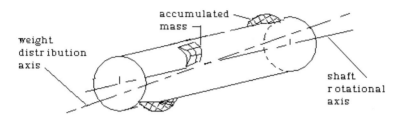

**Fig. 2.10**  Dynamic unbalanced rotor

### 2.6.2.3   Dynamic Unbalance Rotor

It is shown in Fig. 2.10. If this fault occurs then shaft rotational axis and weight distribution axis of rotor do not coincide. It is the combination of coupling unbalance and static unbalance.

The main causes of rotor mass unbalance in an induction motor can be mentioned, pointwise, as follows:

- manufacturing defect
- internal misalignment or shaft bending
- it may occur after an extended period of operation, for nonsymmetrical addition or subtraction of mass around the center of rotation of rotor.

## 2.6.3   Effect of Rotor Mass Unbalance

If in an induction motor rotor mass unbalance occurs, its effect will be as follows:

- Mass unbalance produces dynamic eccentricity which results in oscillation in the air gap length.
- Oscillation in the air gap length causes variation in air gap flux density, and hence variation in induced voltage in the winding.

- Induced voltage causes current whose frequencies are determined by the frequency of the air gap flux density harmonics. The stator current harmonics [21, 22] is given by

$$f_{ubm} = f \left[ \frac{k(1-s)}{p} \right] + 1 \qquad (2.2)$$

where $f$ is the supply frequency, $s$ is the slip of the motor, $p$ is the number of pole pair, and $k$ is an integer.

## 2.7   Bearing Fault

Two sets of bearings are placed at both the ends of the rotor of an induction motor to support the rotating shaft. They held the rotor in place and help it to rotate freely by decreasing the frictions. Each bearing consists of an inner and an outer ring called races and a set of rolling elements called balls in between these two races. Normally, in case of motor, inner race is attached to the shaft and load is transmitted through the rotating balls—this decreases the friction. Using lubricant (oil or grease) in between the races friction is further decreased. Figure 2.11 shows a typical ball bearing and Fig. 2.12 shows a dissected ball bearing.

Any physical damage of the inner race or in the outer race or on the surface of the balls is termed as bearing fault. In terms of induction motor failure, bearing is the weakest component of an induction motor. It is the single largest cause of fault in induction motor. As per the study of IEEE and EPRI, given in Table 2.3, 41–42 % of induction motor faults are due to bearing failure [7, 9].

**Fig. 2.11** Ball bearing

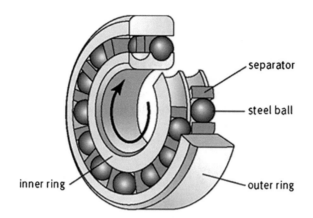

**Fig. 2.12** Ball bearing (dissected)

**Causes and effects of bearing failure:**

1. Excessive loads, tight fits, and excessive temperature rise: all of these can anneal the two races and ball materials. They can also degrade, even destroy, the lubricant. If the load exceeds the elastic limit of the bearing material, brinelling occurs.
2. Fatigue failure: this is due to long run of the bearings. It causes fracture and subsequently removal of small discrete particles of materials from the surfaces of races or balls. This type of bearing failure is progressive, that is, if once initiated will spread when further operation of bearings takes place. For this bearing failure, vibration and noise level of motor will increase [23].
3. Corrosion: this results if bearings are exposed to corrosive fluids (acids, etc.) or corrosive atmosphere. If lubricants deteriorate or the bearings are handled carelessly during installation, then also corrosion of bearings may take place [23]. Early fatigue failure may creep in due to corrosion.
4. Contamination: it is one of the leading factors of bearing failure. Lubricants get contaminated by dirt and other foreign particles which are most often present in industrial environment. High vibration and wear are the effects of contamination.
5. Lubricant failure: for restricted flow of lubricant or excessive temperature this takes place. It degrades the property of the lubricant for which excessive wear of balls and races takes place which results in overheating. If bearing temperature gets too high, grease (the lubricant) melts and runs out of bearing. Discolored balls and ball tracks are the symptoms of lubricant failure.
6. Misalignment of bearings: for this, wear in the surfaces of balls and races takes place which results in rise in temperature of the bearings.

It is observed that for any of the bearing failures, normally friction increases which causes rise in temperature of the bearings and increase in vibration of the concerned machine. For this, bearing temperature and vibration can provide useful information regarding bearing condition and hence machine health [23, 24].

## 2.8  Stator Fault

Stator of an induction motor is subjected [25] to various stresses such as mechanical, electrical, thermal, and environmental [18]. Depending upon the severity of these stresses stator faults may occur. If for a well-designed motor operations and maintenance are done properly, then these stresses remain under control. The stator faults can be classified as (i) faults in laminations and frame of stator and (ii) faults in stator winding. Out of these the second one is the most common stator fault. As per the study of IEEE and EPRI, given in Table 2.3, 28–36 % of induction motor faults are stator winding fault [7, 9]. Majority of these faults are due to a combination of above stresses.

### 2.8.1  Stator Winding Fault

This fault is due to failure of insulation of the stator winding. It is mainly termed as inter-turn short-circuit fault. Different types of stator winding faults are (i) short circuit between two turns of same phase—called turn-to-turn fault, (ii) short circuit between two coils of same phase—called coil to coil fault, (iii) short circuit between turns of two phases—called phase to phase fault, (iv) short circuit between turns of all three phases, (v) short circuit between winding conductors and the stator core—called coil to ground fault, and (vi) open-circuit fault when winding gets break. Different types of stator winding faults are shown in Fig. 2.13. Short-circuit winding fault shows up when total or a partial of the stator windings get shorted. Open-circuit fault shows up when total or a partial of the stator windings get disconnected and no current flows in that phase/line (Figs. 2.14 and 2.15).

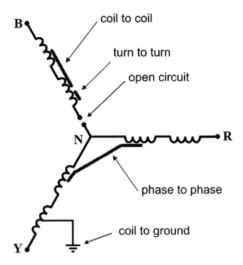

**Fig. 2.13**  Star-connected stator showing different types of stator winding fault

**Fig. 2.14** Photograph of damage stator winding

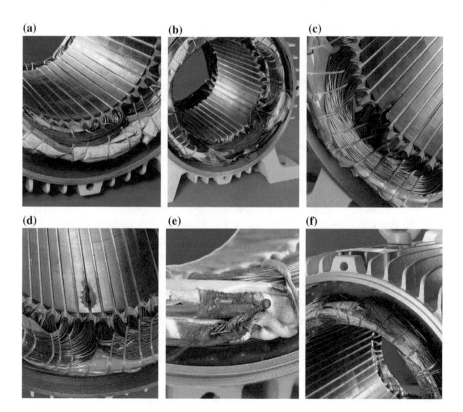

**Fig. 2.15** Typical insulation damage leading to inter-turn short circuit of the stator windings in three-phase induction motors. **a** Inter-turn short circuits between turns of the same phase. **b** Winding short circuited. **c** Short circuits between winding and stator core at the end of the stator slot. **d** Short circuits between winding and stator core in the middle of the stator slot. **e** Short circuit at the leads. **f** Short circuit between phases

## 2.8.2  Causes and Effects of Stator Winding Faults

(i) **Mechanical Stresses**—these are due to movement of stator coil and rotor striking the stator [25]. Coil movement which is due to the stator current (as force is proportional to the square of the current [26]) may loosen the top sticks and also may cause damage to the copper conductor and its insulation. Rotor may strike the stator due to rotor-to-stator misalignment or due to shaft deflection or due to bearing failure and if strikes then the striking force will cause the stator laminations to puncture the coil insulation resulting coil to ground fault. High mechanical vibration may disconnect the stator winding producing the open-circuit fault [27].

(ii) **Electrical Stresses**—these are mainly due to the supply voltage transient. This transient arises due to different faults (like line-to-line, line-to-ground, or three-phase fault), due to lightning, opening, or closing of circuit breakers or due to variable frequency drives [25]. This transient voltage reduces life of stator winding and in severe case may cause turn-to-turn or turn-to-ground fault.

(iii) **Thermal stresses**—these are mainly due to thermal overloading and are the main reason, among the other possible causes, for deterioration of the insulation of the stator winding. Thermal stress happens due to over current flowing due to sustained overload or fault, higher ambient temperature, obstructed ventilation, unbalanced supply voltage, etc. [25]. A thumb rule is there which states that winding temperature will increase by 25 % in the phase having the highest current if there is a voltage unbalance of 3.5 % per phase [18]. Winding temperature will also increase if within a short span of time a number of starts and stops are made in the motor. What may be the reason, if winding temperature increases and the motor is operated over its temperature limit, the best insulation may also fail quickly. The thumb rule, in this regard, states that for every 10 °C increase in temperature above the stator winding temperature limit, the insulation life is reduced by 50 % [28, 18]. Table 2.4 shows the effect of rise of temperature above ambient on the insulation of winding [18].

(iv) **Environmental stresses**—these stresses may arise if the motor operates in a hostile environment with too hot or too cold or too humid. The presence of foreign material can contaminate insulation of stator winding and also may reduce the rate of heat dissipation from the motor [29], resulting reduction in

**Table 2.4** Effect of rise of temperature

| Ambient in °C | Insulation life in hours |
| --- | --- |
| 30 | 250,000 |
| 40 | 125,000 |
| 50 | 60,000 |
| 60 | 30,000 |

insulation life. Air flow should be free where the motor is situated, otherwise the heat generated in the rotor and stator will increase the winding temperature which will reduce the life of insulation.

## 2.9  Single Phasing Fault

This is a power supply-related electrical fault in case of an induction motor. For a three-phase motor when one of the phases gets lost then the condition is known as single phasing.

### 2.9.1  Causes of Single Phasing Fault

Single phasing fault in an induction motor may be due to

- A downed line or a blown fuse of the utility system.
- Due to an equipment failure of the supply system.
- Due to short circuit in one phase of the star-connected or delta-connected motor.

### 2.9.2  Effects of Single Phasing Fault

Effects of single phasing fault are as follows:

- For single phasing fault motor windings get over heated, primarily due to flow of negative sequence current.
- If during running condition of the motor single phasing fault occurs motor continues to run due to the torque produced by the remaining two phases and this torque is produced as per the demand by the load—as a result healthy phases may be over loaded and hence over heated resulting in critical damage to the motor itself.
- A three-phase motor will not start if a single phasing fault already persists in the supply line.

## 2.10  Crawling

It is an electromechanical fault of an induction motor. When an induction motor, though the full-load supply is provided, does not accelerate but runs at a speed nearly one-seventh of its synchronous speed, the phenomenon is known as crawling of the motor.

### 2.10.1  General Description

The air-gap flux in between stator and rotor of an induction motor is not purely sinusoidal because it contains some odd harmonics. Due to these harmonics, unwanted torque is developed. The flux due to third harmonics and its multiples produced by each of the three phases differs in time phase by 120° and hence neutralize each other. For this reason, harmonics present in air-gap flux are normally 5th, 7th, 11th, etc.

The fundamental air-gap flux rotates at synchronous speed given by $N_s = 120f/P$ rpm where f is the supply frequency and $P$ is the number of poles. However, harmonic fluxes rotate at $N_s/k$ rpm speed ($k$ denotes the order of the harmonics), in the same direction of the fundamental except the 5th harmonic. Flux due to 5th harmonic rotates in opposite direction to the fundamental flux. Magnitudes of 11th and higher order harmonics being very small 5th and 7th harmonics are the most important and predominant harmonics.

Like fundamental flux these two harmonic fluxes also produce torque. Thus total motor torque has three components—(i) fundamental torque rotating at synchronous speed $N_s$ (ii) 5th harmonic torque rotating at speed $N_s/5$ in the opposite direction of fundamental, and (iii) 7th harmonic torque rotating at speed $N_s/7$ in the same direction of fundamental. Thus 5th harmonic torque produces a breaking action whose magnitude is very small and hence can be neglected; consequently, the resultant torque can be taken as the sum of the fundamental torque and the 7th harmonic torque as shown in Fig. 2.16. The 7th harmonic torque has value zero at one-seventh of the synchronous speed. The resultant torque shows a dip near slip 6/7, which is more significant because torque here decreases with increase in speed. The motor under loaded condition will not accelerate up to its normal speed but will remain running at a speed nearly one-seventh of the synchronous speed. This phenomenon is called crawling of the induction motor. It is predominant in the squirrel-cage type induction motor. By proper selection of the number of stator and rotor slots, the crawling effect can be reduced.

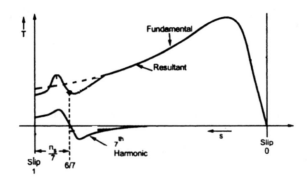

**Fig. 2.16** Torque–slip curve showing resultant of fundamental and 7th harmonic torque

## 2.10.2  Causes of Crawling

Crawling is caused by the 7th harmonic. The 7th harmonic is introduced due to abnormal magneto motive force. Another reason is high harmonic content in the power supply to the motor.

## 2.10.3  Effects of Crawling

Following are the effects of crawling:

- Motor under loaded condition will not accelerate up to its normal speed.
- Loaded motor will remain running at a speed nearly one-seventh of the synchronous speed.
- There will be much higher stator current.
- Motor vibration and noise will be high.

## 2.11  Over Voltage, Under Voltage, Overload, and Blocked Rotor

Over and under voltages occur due to change of voltage level at supply end. Over voltage causes stress on insulation, whereas under voltage causes excessive line current increasing temperature of the winding. These faults are normally detected by over/under voltage relays. Overload occurs due to increase of mechanical load above the rating of the motor. At excessive mechanical load, rotor fails to rotate and gets blocked. This situation is equivalent to short circuit. Normally, overload and blocked rotor are protected by over current relay or simply fuse.

## 2.12  Condition Monitoring and Its Necessity

Induction motors are the main workhorse of industrial prime movers due to their ruggedness, low cost, low maintenance, reasonably small size, reasonably high efficiency, and operating with an easily available power supply. About 50 % of the total generated power of a nation is consumed by these induction motors [30]. This statistics gives an idea regarding the use of huge number of induction motors, but they have some limitations in their operating conditions. If these conditions exceed then some premature failure may occur in stator or/and rotor. This failure, in many applications in industry, may shut down, even, the entire industrial process resulting loss of production time and money. Hence, it is an important issue to avoid any kind of failure of induction motor. Operators and technicians of induction motors are under continual pressure to prevent unscheduled downtime and also to reduce maintenance cost of motors.

Maintenance of electrical motors can be done in three forms: breakdown maintenance, fixed-time maintenance, and condition-based maintenance. In breakdown maintenance, the strategy is 'run the motor until it fails' which means maintenance action is taken only when the motor gets break down. In this case though the motor may run comparatively for a long time before the maintenance is done but when break down occurs it is necessary to replace the entire machine which is much costlier compared to replacing or repairing the faulty parts of the motor. Also it causes loss of productivity due to downtime. In fixed-time maintenance, motor is required to stop for inspection which causes long downtime. Also trained and experienced technical persons are required to recognize each and every fault correctly. All these necessitate the condition-based maintenance of the motor. In this form of maintenance, motor is allowed to run normally and action is taken at the very first sign of an incipient fault. There are a number of works [31–37] on this online condition monitoring of induction motor.

In condition monitoring, when a fault has been identified, sufficient data is required for the plant operator for the best possible decision making on the correct course of action. If data is insufficient there remains the chance for wrong diagnosis of fault which leads to inappropriate replacement of components, and if the root of the problem is not identified properly, the replacement or any other action taken already will succumb to the same fate.

In condition monitoring, signals from the concerned motor are continuously fed to the data acquisition system and the health of the motor is continuously evaluated during its operation for which it is also referred as online condition monitoring of motor, and hence it is possible to identify the faults even while they are developing. The operator/technician can take preparation for the preventive maintenance and can arrange for necessary spare parts, in advance, for repairing. Thus condition monitoring can optimize maintenance schedule and minimize motors downtime [9] and thereby increase the reliability of the motor. Advantages of using condition monitoring can be mentioned pointwise as below:

- Can predict the motor failure.
- Can optimize the maintenance of the motor.
- Can reduce the maintenance cost.
- Can reduce downtime of the machine.
- Can improve the reliability of the motor.

# References

1. Krause PC (1986) Analysis of electric machinery. Mc-Graw Hill, New York
2. Sen PC (1989) Principles of electric machines and power electronics. Willey, New York
3. Say MG (2002) The performance and design of alternating current machines. M/S Pitman, London. ISBN 81-239-1027-4

4. Kothari DP, Nagrath IJ (2010) Electric machines. Tata McGraw Hill Education Pvt. Ltd., New Delhi. ISBN-13 978-0-07-069967-0
5. Motor Reliability Working Group (1985) Report of large motor reliability survey of industrial and commercial installations Part I, and II. IEEE Trans Ind Appl IA-21(4):853–872
6. Thorsen OV, Dalva M (1995) A survey of faults on induction motors in offshore oil industry, petrochemical industry, gas terminals, and oil refineries. IEEE Trans Ind Appl 31(5):1186–1196
7. Singh GK, Al Kazzaz SAS (2003) Induction machine drive condition monitoring and diagnostic research—a survey. Electr Power Syst Res 64(2):145–158
8. IEEE recommended practice for the design of reliable industrial and commercial power systems. IEEE Standard 493–1997 [IEEE Gold Book]
9. Allbrecht PF, Appiarius JC, McCoy RM, Owen EL (1986) Assessment of the reliability of motors in utility applications—updated. IEEE Trans Energy Convers EC-1(1):39–46
10. Bonnett AH, Soukup GC (1988) Analysis of rotor failures in squirrel-cage induction motors. IEEE Trans Ind Appl 24:1124–1130
11. Su H, Chong KT, Kumar RR (2011) Vibration signal analysis for electrical fault detection of induction machine using neural networks. Neural Comput Appl 20(2):183–194, Springer
12. Vas P (1993) Parameter estimation, condition monitoring and diagnosis of electrical machines. Clarendon Press, Oxford
13. Tavner PJ, Penman J (1987) Condition monitoring of electrical machines. Research Studies Press Ltd., Hertfordshire, England. ISBN 0863800610
14. Bonnet AH, Soukup GC (1992) Cause and analysis of stator and rotor failures in three phase squirrel cage induction motors. IEEE Trans Ind Appl 28(4):921–937
15. Deleroi W (1984) Broken bars in squirrel cage rotor of an induction motor-part I: description by superimposed fault currents. Arch Elektrotech 67:91–99
16. Filippetti F, Franceschini G, Tassoni C, Vas P (1998) AI techniques in induction machines diagnosis including the speed ripple effect. IEEE Trans Ind Appl 34:98–108
17. Bellini A, Concari C, Franceschini G, Lorenzani E, Tassoni C, Toscani A (2006) Thorough understanding and experimental validation of current sideband components in induction machines rotor monitoring. In: IECON 2006-32nd annual conference on IEEE industrial electronics, pp 4957–4962
18. Bonnett Austin H, Soukup GC (1992) Cause and analysis of stator and rotor failures in three phase squirrel-cage induction motors. IEEE Trans Ind Appl 28(4):921–937
19. Dorrell DG, Thomson WT, Roach S (1997) Analysis of air-gap flux, current and vibration signals as function of a combination of static and dynamic eccentricity in 3-phase induction motors. IEEE Trans Ind Appl 33:24–34
20. Bradford M (1968) Unbalanced magnetic pull in a 6-pole induction motor. IEEE Proc Electr Eng 115(11):1619–1627
21. M'hamed D, Cardoso AJM (2008) Air gap eccentricity fault diagnosis in three phase induction motor by the complex apparent power signature analysis. IEEE Trans Ind Electr 55(3):1404
22. Hwang DH, Lee KC, Lee JH, Kang DS, Lee JH, Choi KH, Kang S et al (2005) Analysis of a three phase induction motor under eccentricity condition. In: 31st annual conference of IEEE industrial electronics society, IECON2005, pp 6–10
23. Eschmann P, Hasbargen L, Weigand K (1958) Ball and roller bearings: their theory, design and application. K. G. Heyden, London
24. Schoen RR, Habetler TG, Kamran F, Bartheld RG (1995) Motor bearing damage detection using stator current monitoring. IEEE Trans Ind Appl 31(6):1274–1279
25. Siddique A, Yadava GS, Singh B (2005) A review of stator fault monitoring techniques of induction motors. IEEE Trans Energy Convers 20(1):106–114
26. Lee SB, Tallam RM, Habetler TG (2003) A robust on-line turn-fault detection technique for induction machines based on monitoring the sequence component impedance matrix. IEEE Trans Power Electr 18(3):865–872

27. Aguayo J, Claudio A, Vela LG, Lesecq S, Barraud A (2003) Stator winding fault detection for an induction motor drive using actuator as sensor principle. IEEE xplore 0-7803-7754-0/03@2003 IEEE
28. Lipo TA (2004) Introduction of AC machine design, 2nd edn. Wisconsin Power Electronics Research Center, Madison
29. Cash MA (1998) "Detection of turn faults arising from insulation failure in the stator windings of AC machines", Doctoral Dissertation. Department of Electrical and Computer Engineering, Georgia Institute of Technology, USA
30. Thomson WT, Gilmore RJ (2003) Motor current signature analysis to detect faults in induction motor drives-fundamentals, data interpretation, and industrial case histories. In: Proceedings of the thirty-second turbomachinery symposium, pp 145–156
31. Jung JH, Lee JJ, Kwon BH (2006) Online diagnosis of induction motors using MCSA. IEEE Trans Ind Electron 53(6):1842–1852
32. Tavner PJ (2008) Review of condition monitoring of rotating electrical machines. IET Electr Power Appl 2(4), 215
33. Nandi S, Toliyat HA (1999) Condition monitoring and fault diagnosis of electrical machines—a review. In: Proceedings 34th annual meeting of IEEE industrial applications society, pp 197–204
34. Ahmed I, Supangat R, Grieger J, Ertugrul N, Soong WL (2004) A baseline study for online condition monitoring of induction machines. In: Australian Universities power engineering conference (AUPEC), Brisbane, Australia
35. Thomson WT, Rankin D, Dorrell DG (1999) On-line current monitoring to diagnose air gap eccentricity in large three-phase induction motors—industrial case histories to verify the predictions. IEEE Trans Energy Convers 14(4):1372–1378
36. Thomson WT, Barbour A (1998) On-line current monitoring and application of a Finite Element method to predict the level of air gap eccentricity in 3-Phase induction motor. IEEE Trans Energy Convers 13(4):347–357
37. Wolbank TM, Macheiner PE (2007) Adjustment, measurement and on-line detection of air gap asymmetry in AC machines. IEEE, Vienna, Austria
38. Benbouzid MEH (2000) A review of Induction motor Signature analysis as a medium for Fault detection. IEEE Trans Ind Electron 47(5):984–993
39. Gaeid KS, Mohamed HAF (2010) Diagnosis and fault tolerant control of the induction motors techniques a review. Aust J Basic Appl Sci 4(2):227–246. ISSN 1991-8178

# Chapter 3
# Analytical Tools for Motor Fault Diagnosis

**Abstract** This chapter deals with different approaches useful for motor fault diagnosis. It mentions different methods like thermal analysis, acoustics analysis, vibration analysis, etc. Then, mathematical tools used for signal processing and analysis of steady as well as transient currents are discussed. It covers fast Fourier transform (FFT), wavelet transform (WT), discrete wavelet transform (DWT), Concordia, radar analysis, etc.

**Keywords** Acoustics analysis · Concordia · Discrete wavelet transform (DWT) · Fast Fourier transform (FFT) · Radar analysis · Thermal analysis · Vibration analysis · Wavelet transform (WT)

### Chapter Outcome

After completion of this chapter, readers will be able to gather knowledge and information regarding the following areas:

- Different approaches used for motor fault diagnosis
- Different mathematical tools for motor fault diagnosis
- Fast Fourier Transform
- Discrete Fourier Transform
- Wavelet transform
- Skewness
- Kurtosis
- CMS Rule Set
- Concordia and Radar assessment.

## 3.1 Introduction

Early detection of fault is a challenge and a lot of research works are going on for long to develop new analytical tools. For fault detection of induction motor, different techniques like motor current signature analysis (MCSA), vibration analysis, thermal analysis, etc., are used. Research has proved that various signal processing

© Springer Science+Business Media Singapore 2016
S. Karmakar et al., *Induction Motor Fault Diagnosis*,
Power Systems, DOI 10.1007/978-981-10-0624-1_3

tools can be used successfully under different running conditions such as starting or steady under load or no-load of the motor for fault diagnosis. Each tool has some advantages or limitations in a particular condition of the motor. In this chapter FFT, Wavelet Transform (WT), Hilbert Transform, Radar analysis of current Concordia and feature pattern extraction method will be discussed.

## 3.2  Existing Techniques for Fault Analysis of Induction Motor

In condition monitoring, signals obtained from the motor are analyzed continuously. Based on these signals, researchers have used different techniques for diagnosis of faults of induction motors.

### 3.2.1  Thermal Analysis

By thermal analysis fault detection of an induction motor is performed normally by measuring the change in temperature of the motor. Thermal analysis can be used to detect bearing fault and turn-to-turn stator winding fault of an induction motor. In case of bearing fault, friction increases which in turn increases the temperature of the motor. In case of turn-to-turn fault temperature increases in the fault region, as a result detection of the fault would take time. Thermal analyzing this change in temperature motor fault can be predicted.

A number of works [1–5] have been done on thermal analysis of motor. In [1], to identify the cause of failure and hence to improve the reliability of an induction motor transient thermal behavior of the motor has been studied using heat transfer coefficient. Overall behavior of the motor temperature has been studied in [2] during the transient and steady-state operation by thermal circuit model where calculation of the temperature distribution is proposed in the radial direction, i.e., from the shaft to the frame surface of the motor. In [3], thermal modeling of electric machines has been done in two ways (i) lumped parameter thermal model and (ii) finite element analysis-based model. In lumped parameter thermal model one electric machine is assumed to be made of several thermally homogenous lumped bodies.

### 3.2.2  Chemical Analysis

Normally, chemical analysis of lubricants of motor are performed to detect the fault in an induction motor. Bearing fault of an induction motor can be identified by this technique. Black deposits in loaded area of pad suggest chemical attack. By X-ray these deposits can be identified. Bearing of copper alloys are more prone to attack than bearings made of tin-rich white metal. Chemical analysis is used only for big motors and not for small motors [6].

### 3.2.3 *Acoustic Analysis*

This is performed by measuring and analyzing the acoustic noise spectrum [7] generated by the motor. If faults like bearing fault, air gap eccentricity fault occur, then these spectrum changes. For this reason, acoustic noise can be used for fault detection in induction motor. Ellison and Young [8] have reported in their paper the effects of rotor eccentricity on acoustic noise from induction machines. Lee et al. [9] has also worked on acoustic analysis technique. They have used this method for early detection of incipient fault. In their work, for the purpose of acoustic analysis, an ultrasonic wave has been introduced into the stator and the conductor has been used as a waveguide. Tahori et al. have demanded in their paper [10] that in the field of gearbox failure detection, acoustic analysis is a new trend.

However in a noisy environment like a plant where a number of other machines are working, acoustic analysis may not be practical [11].

### 3.2.4 *Torque Analysis*

Most of the faults in an induction motor produce harmonics of specific frequencies in the air gap. But this air gap torque cannot be measured directly. Also the air gap torque is different from the torque measured at the shaft of the motor. In terms of motor terminal parameters which are measurable, air gap torque in Newton-meter can be expressed as [7]

Torque

$$
= \frac{p}{\sqrt{2}} \left[ (i_a - i_b) \int [V_{ca} - R(i_c - i_a) \mathrm{d}t] \\
- (i_c - i_a) \int [V_{ab} - R(i_a - i_b) \mathrm{d}t] \right]
\tag{3.1}
$$

where $i_a$, $i_b$ and $i_c$ are line currents of the induction motor, $V_{ab}$ and $V_{ca}$ are line-to-line voltages, $p$ is number of pole pairs, and $R$ is half of the line-to-line resistance.

Now frequencies of major harmonic torques associated with certain faults in induction motors are as follows [7]:

- Under normal operation, angular frequency of torque = 0
- Under single-phasing stator, angular frequency of torque = $-2\omega_s$
- Under single-phasing rotor, angular frequency of torque = $2s\omega_s$

where $\omega_s$ is the supply frequency in rad/s and $s$ is the slip of the induction motor.

From the above discussions, it can be concluded that by analyzing the harmonics present in the air gap torque, fault in the induction motor can be diagnosed.

A number of works [11–14] have been performed to diagnose induction motor fault by analyzing torque and air gap frequencies. In [11] air gap torque profile has been analyzed to determine the condition (healthy or faulty) of the motor. Here, air gap torque normalization method has been introduced and for it motor terminal operating voltages and currents are taken as data to classify the motor fault. Stopa et al. [12] have shown in their paper that load torque signature analysis (LTSA) can be an alternative to MCSA. In this work, analyzing the influence of electrical and mechanical parameters on LTSA, mathematical relations have been derived from which the nature of the frequency responses is obtained. Arabaci et al. [14] have presented in their work the effects of rotor faults on torque–speed curve of induction motor. They collected data during motor starting under no-load condition for healthy and faulty motors with one, two, and three broken rotor bar and motor with broken end ring. Analyzing the data using FFT and STFT, researchers demand that torque–speed curves are affected proportionately depending on the size of the rotor faults.

### 3.2.5 Induced Voltage Analysis

By analyzing the induced voltage along the shaft of a machine, fault can be identified. This voltage is induced due to degradation of insulation of stator winding. Normally, this induced voltage is very small and is measurable when a significant amount of damage in the stator winding occurs [15]. For this reason, induced voltage technique is not widely used. In Cash et al. [16] have given an idea of prediction of insulation failure using line-to-neutral voltage.

### 3.2.6 Partial Discharge Analysis

Due to degradation of winding insulation a small electrical discharge occurs—this is referred as partial discharge. A deteriorated insulation of winding may have a partial discharge activity of 30 times or more than a winding in good condition [17]. To detect motor failures a number of research works [18–21] have been performed by analyzing this partial discharge. In fact in a high voltage machine time taken for failure being very small, online partial discharge monitoring is used to assess the insulation health of stator winding. It has been proved by the researchers [18, 19] that partial discharge analysis can identify degradation in insulation prior to complete breakdown in a high voltage machine. This technique is being widely used in industry and its validity has been verified in [20] by Stone et al. The stator winding maintenance can be monitored with online partial discharge analyzer (PDA) test [21].

## 3.2.7 Vibration Analysis

Induction motor generates vibration and this vibration depends on the radial forces due to the air gap flux. Air gap flux distribution depends on the resultant m.m.f. wave and permeance. Now the resultant m.m.f. wave varies if rotor asymmetries occur and permeance varies with the variation of the air gap. All these variations occur if any fault especially mechanical faults occur in the motor. Hence analyzing the motor vibration it is possible to predict the faults and various types of asymmetries in the motor [22]. By vibration analysis bearing fault, gear fault, unbalanced rotor fault, fault due to rotor eccentricities can be identified [23].

In vibration analysis, the oscillation force which is imparted by the motor is sensed by a sensor. This force is linearly related with the oscillation acceleration. For vibration analysis, researchers often use both the vibration acceleration and the vibration velocity in restricted low-frequency ranges [24]. In most of the vibration analysis vibration acceleration is measured using a piezoelectric transducer which works based on the piezoelectric effect (production of electricity from mechanical stress). These piezoelectric transducers are quite costly which limits the use of vibration analysis technique for fault detection especially in small motors.

Literature survey reveals that a number of researches [25–33] are performed to diagnose motor fault by analyzing vibration signal. In [25] vibration signal has been analyzed using neural network for electrical fault detection of an induction machine by Su et al. Widodo et al. in [26] have used Independent Component Analysis (ICA) and Support Vector Machine (SVM) on data sets of both vibration signals and stator current signals to detect and diagnose faults in induction motors. In [27] motor vibration signals under loaded and no-load condition have been analyzed to detect broken rotor bar fault using MEMS accelerometer. Performing spectral analysis of motor vibration by FFT authors have found twice slip frequency components around fundamental frequency through which they have concluded the fault as broken rotor bar fault. In Wang et al. [29] have modeled a 2.2 kW induction motor. They have analyzed the effect of the stator, end shields, outer casing and supports on the behavior of the overall vibration of the system and have concluded that they affect the vibration of the system. Tsypkin in his work [31] has shown that for induction motor, vibration analysis is a very efficient and convenient tool to diagnose mechanical problems like bearing fault, mechanical unbalance, structural resonance, foundation problems, etc., and also combination of mechanical and electrical problems like winding damage, voltage waveform distortion, etc. In [33] researchers have worked to detect bearing fault by both MCSA and vibration analysis using FFT algorithm and relationship between them is verified. It is demanded that identification of fault frequency by MCSA is more difficult than vibration analysis. Here it is indicated that reinstalling a faulty bearing may alter the characteristic frequency of the motor. In Schoen et al. [34] have studied the relationship between vibration frequencies and current frequencies. In doing so, authors have investigated the detection of motor bearing fault using stator current for different bearing faults. Authors have concluded that the bearing fault can be identified by the use of stator current signature.

### 3.2.8   Current Analysis

In this technique, stator current is analyzed. Now it is seen that in all most all the techniques discussed above, to get the parameter to be analyzed, transducers are required to be fitted in or around the motor, which may interrupt its operation [35] besides the cost. But stator current may be obtained even without any extra device if collected from the already installed devices for metering purpose or may be easily available from the protective devices of overcurrent, ground current, etc., of the studied motor. In [36] it is concluded that MCSA is a sensorless detection method which can be implemented without any extra hardware. Stator current can be measured online, meaning that the data for current analysis technique is achievable at all time when the motor is running. For these reasons current signature has become a practical parameter for detecting faults of squirrel cage induction motor. Most of the mechanical and electrical faults those may arise in an induction motor are detectable by this current analysis technique. This is also termed as MCSA, abbreviated as MCSA, or sometimes simply as CSA—current signature analysis.

Current signature analysis technique has mainly three steps as shown in Fig. 3.1 where each of the steps is shown as a block. The steps are (i) data acquisition, (ii) feature extraction, and (iii) fault assessment.

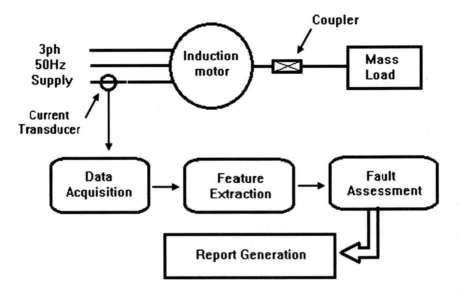

**Fig. 3.1** Block diagram showing current signature analysis technique for fault detection

(i) **Data acquisition**: In this step of current signature analysis, motor stator current is captured with the help of current transducer either a current transformer or a Hall sensor. Normally, one of the phases is monitored but sometimes for more accuracy all the three phases are monitored with the help of three identical transducers. The captured data is first digitized by sampling the same. In some works this data is filtered to remove undesired components of frequencies. This is called preprocessing. Sampled data is stored for further use like feature extraction.

(ii) **Feature extraction**: In the second step of current signature analysis the digitized, preprocessed, stored data is processed for feature extraction. Here comes the term digital signal processing (DSP). Data is processed in time domain or in frequency domain or in both time–frequency domains. For processing purpose, different tools like fast Fourier transform (FFT), discrete Fourier transform (DFT), short-time Fourier transform (STFT), Wavelet Transform (WT), Hilbert transform (HT), support vector machine (SVM), Park's vector approach, feature pattern extraction method (FPEM), etc., are used. Also artificial intelligence-based techniques like neural network (NN), fuzzy, neuro-fuzzy, etc., are used for data processing to extract features. The details of these signal processing tools will be discussed in subsequent sections. All the above tools analyze the current signature to separate the constituent frequencies. From this generated spectrum, purpose of CSA is to detect the presence of harmonic components which are generated due to faults in the particular motor.

(iii) **Fault assessment**: In this level fault is identified and its severity is calculated. Normally, this is done by the comparison of the frequency spectrum of the studied motor with that of healthy motor for which accurate knowledge of the motor being studied is required [37]. A typical two-level knowledge-based system is shown in [38] for the diagnosis of rotor fault.

MCSA technique is applicable for detection of a number of faults like (i) rotor broken bar fault, (ii) air gap eccentricity or mass unbalance fault, (iii) stator winding fault. Stator current analysis is noninvasive, precise, low cost, easy, and efficient technique. It can provide information as the vibration analysis can provide [39]. Moreover sensor required for all other analysis techniques being costlier this MCSA is gaining much importance and researches around this are growing fast. Using MCSA there have been ample research works [6, 17, 36, 40–48].

Data for MCSA may be either steady-state current of motor or the starting current which is transient in nature. Steady-state current is easily available and may be captured from a running motor at any stage. A number of researchers have considered steady-state current as data for current signature analysis in their works [17, 36, 40–47, 49–52]. Schoen et al. [17] in their work have detected the bearing damage of motor by monitoring the stator current. Online condition monitoring have been performed by researchers in [41, 42] using MCSA. Air gap eccentricity fault has been diagnosed by online current monitoring in [43, 44] by Thomson et al. da Silva et al. have worked with stator current envelopes for MCSA in [45] to diagnose broken bar rotor fault and short-circuit stator fault in an induction machine.

*Disadvantage of steady-state data*: Data collection during steady state of the motor is that if the speed of the motor changes during the sampling time, then the extracted spectrum will get blurred [37]. Then the FFT tool will not work properly due to this nonstationary nature of the motor. Also the amplitude of the steady-state current depends on load connected to the motor. For this MCSA using steady-state current not applicable where there is any chance of variation in system load [53].

This deficiency of MCSA is overcome by considering the starting transient current of motor as the required signal, because this signal is less affected by the motor loading and hence signal may be captured even when the motor is running at no-load or light load. Also another advantage is that starting current being about 7–8 times of the steady-state current, even though test is performed on small-sized motor variation in current due to rotor fault will be much more evident [54]. A number of research works [6, 52, 55–61] have been performed by analyzing the transient current of induction motors to detect their faults. In [6] Jordi Cusido et al. have shown the method of detection of motor fault by analyzing the transient current of the motor using the WT as signal processing tool. In Ye et al. [62] have worked on mechanical faults of induction motor by wavelet packet decomposition of current signature.

## 3.3  Signal Processing Tools for Fault Analysis

For MCSA, the data which is either steady-state current or starting current are analyzed by different signal processing tools which are discussed below.

### 3.3.1  Fast Fourier Transform (FFT)

Fourier transform (FT) converts a signal into frequency domain from time domain and inverse Fourier transform converts a signal into time domain from frequency domain. In doing so number of computations needed is $2N^2$ where $N$ is number of data. FFT also do the same, i.e., converts a signal into frequency domain from time domain and vice versa. But in this case the number of computations needed is $2N \log_2 N$. Now the difference between $2N^2$ and $2N \log_2 N$ is immense. Thus number of computations being small compared to the Fourier transform this technique can perform the transformation rapidly hence it is named as FFT. For long data sets (1000 or more) the difference in speed for performing the transformation becomes enormous hence FFT has become a fundamental, practical, and convenient tool for DSP system. It is widely used in science, engineering, communication, metallurgy, applied mechanics, biomedical engineering, radar etc.

FFT is a steady-state analysis, i.e., this analysis is applied on the current signal taken from the motor when it is running in steady state. FFT gives different frequency components present in the signal. Over the last few years there has been a substantial amount of research work based on FFT [41, 42, 44, 63–66]. Researchers have proved that this tool can be successfully used for rotor fault detection. Menacer et al. [63] have used, in their work, stator current to analyze rotor bar fault of asynchronous motor. In [65] FFT is used by Marcelo et al. to detect failures of induction motor online. Here researchers have worked on broken rotor bar fault, bearing damage, eccentricity, and short-circuit fault. Exact values of harmonics created due to faults, regardless of sampling time, have been ascertained. Researchers have demanded that the exact extent of the defects and the associated frequencies can be identified by FFT analysis.

There are some limitations in performance of FFT application. Pointwise they can be mentioned as (1) sampling frequency, for application of FFT, must be greater than twice the highest frequency of the signal to be analyzed. Also the window length of data must be an integer multiple of the power supply frequency. If these points are not satisfied then FFT will give inaccurate waveform frequency analysis also aliasing and leakage effects may result in. (2) Power supply frequency leakage which means supply frequency may completely mask the characteristics frequency components of the studied motor specially if the studied motor is of small or medium sized and is running at light load or no-load. It is shown in [66] that in case of broken bar faulted motor characteristic frequencies ($\pm 2fs$) generated are very near to the fundamental frequency ($f$) and their amplitudes are also very small in comparison to the fundamental. Hence detection of fault and determination of fault severity under light load is not possible especially for small motors. (3) Another serious limitation of FFT is that it is inappropriate for the signal whose characteristic changes with time. It is because in transforming to the frequency domain, the time information is lost hence by FFT transformation frequency components localization in any spectrum is not possible. For this reason, in the analysis of starting current of a direct on line (DOL) starting motor, FFT is inappropriate, as this current is transient in nature. Also due to variation of motor load, inertia, torque, supply voltage, or speed oscillation of motor some small harmonics may generate which are similar to the characteristic frequencies for the faulted motor. For this, diagnosis of motor fault by observation of motor current frequencies will be confusing and wrong [56]. This problem was shorted out by the use of a DSP tool called short-t-ime Fourier transform (STFT).

### 3.3.2 Short-Time Fourier Transform (STFT)

The STFT is a Fourier-based transformation normally used to map a signal into two variables, namely time and frequency of which time is discrete and frequency variable is continuous. STFT is capable to analyze transient signal which varies with time.

If in a sequence the time index is fixed then STFT becomes a normal Fourier transform of the sequence. In STFT keeping frequency as fixed quantity interpretation is done as a function of time index. With a particular value of the frequency, interpretation leads to consider the STFT as of linear filtering. Interpretation in terms of linear filtering is useful when only a number of particular frequencies are required to be identified. Then this approach can be used to determine the fundamental frequency and its integer multiple. Another approach is the application of window function to input signal. Advantage of this application is that periodical signal can be analyzed without determining the integer multiple of its periods. There are different types of window functions. Among them Hamming window and Hanning window functions are suitable for harmonic and interharmonic estimation.

STFT uses constant-sized window to analyze all frequencies—this is the limitation of this method. In [67] it is described that this limited window may find it difficult to match the frequency content of the signal which is generally not known prior to the analysis. To overcome, this limited sized window is required to be replaced by a variable-sized window. In WT suitable variable-sized window is used.

### 3.3.3  Wavelet Transform (WT)

WT is an advanced powerful signal processing tool which provides both time and frequency information of a signal by decomposing it into different scales at different levels of resolution through adjusting the time-widths to the frequencies of a single prototype function called mother wavelet. This adjustment, also called dilation, is done in such a way that higher frequency wavelets will be compressed one and those with lower frequency will be stretched. By this way WT overcomes the limitation of constant sized window of the STFT tool. It is suitable for analyzing transient signals which are typically non-periodic containing high-frequency impulses superimposed on the power frequency and its harmonics. The governing equation of WT is shown in (3.2).

$$C(a, b) = \frac{1}{\sqrt{a}} \int\limits_{-\infty}^{\infty} x(t) \Psi\left(\frac{t - b}{a}\right) dt \qquad (3.2)$$

Equation of the WT has two real parameters $a$, the dilation parameter and $b$, the translation parameter which are also called wavelet scale and position, respectively, and $a \neq 0$. Here, $x(t)$ is the signal, $\psi(t)$ is the mother wavelet function. Two types of WT are used—continuous wavelet transform (CWT) and discrete wavelet transform (DWT).

The continuous wavelet transform (CWT) of a continuous time signal $x(t)$ is defined as follows:

$$\text{CWT}(a, b) = \frac{1}{\sqrt{a}} \int\limits_{-\infty}^{\infty} x(t) \Psi_{a,b}^*\left(\frac{t - b}{a}\right) dt \qquad (3.3)$$

Here $\Psi^*(t)$ denotes the complex conjugate of the mother wavelet function and $a$ and $b$ are real quantity and $a \neq 0$. The CWT operates over every possible scale and position, i.e., here continuous dilation and translation occurs.

If the mother wavelet is dilated and translated discretely by selecting $a = a_0^j$ and $b = kb_0a_0^j$ where $a_0$ and $b_0$ are fixed values such that $a_0 > 1$ and $b_0 > 0$ and $j, k$ are positive integers then the mother wavelet becomes discrete. Using this discrete mother wavelet the corresponding WT is termed as DWT. At each level of scaling and for various positions, the correlation between the signal and the wavelet are called wavelet coefficients. Equation for DWT is given in (3.4)

$$\text{DWT}(j,k) = \frac{1}{\sqrt{a_0^j}} \int_{-\infty}^{\infty} x(t)\Psi_{j,k}^* \left( \frac{t - kb_0a_0^j}{a_0^j} \right) \tag{3.4}$$

As DWT uses a specific subset of scale and positional values it is less computationally complex and takes less computation time compared to CWT [52]. Generally, DWT is used for data compression if signal is already sampled and CWT for signal analysis. The DWT is computed by passing the signal successively through low-pass filters and high-pass filters. The high-pass filter coefficients are termed as detail coefficients (CD) and the low-pass filter coefficients are termed as approximate coefficients (CA). Each step of the decomposition of the signal corresponds to a certain resolution. The decomposition can be iterated with successive approximation. Figure 3.2 shows typical wavelet decomposition.

In Fig. 3.2 sequences $h_w$ and $g_w$ are quadrature mirror filter (QMF) bank. The sequence $h_w$ is known as low-pass filter, while $g_w$ is known as high-pass filter. At each scale, the number of the DWT coefficients of the resulting signals is half that of the decomposed signal.

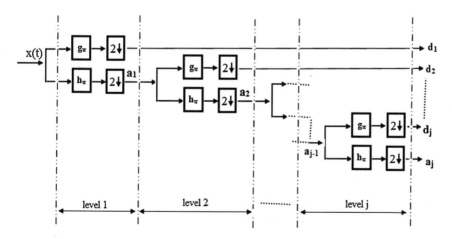

**Fig. 3.2**  Multi-resolution decomposition of discrete wavelet transforms

Now, if the signal $x(t)$ undergoes a $j$th level DWT, then it can be represented approximately as a superposition of scaling functions and wavelets as follows:

$$x(t) = \sum_k a_{j_0,k}\phi_{j_0,k}(t) + \sum_{j \geq j_0} \sum_k d_{j,k}\psi_{j,k}(t) \tag{3.5}$$

where scaling level $j_0$ is the lowest band of the original signal and $j$ includes the signals of successively higher octave frequency bands. $a_{j_0,k}$ are the signal discrete wavelet coefficients at the scaling level $j_0$ and sample $k$, while $d_{j,k}$ are the signal discrete wavelet coefficients at any other level $j$ than the scaling level $j_0$ and sample $k$. The scaling function $\phi_{j_0,k}$ and the wavelet basis $\psi_{j,k}$ are of orthonormal basis. Decomposed wave can be reconstructed as shown in Fig. 3.3 where sequences $h_w$ and $g_w$ are quadrature mirror filter (QMF) bank. The sequence $h_w$ is known as low-pass filter, while $g_w$ is known as high-pass filter. At each scale, the number of the DWT coefficients of the resulting signals is twice that of the reconstructed signal.

Researchers have applied WT both in discrete and continuous form for detection of motor fault by the analysis of the transients using start-up motor current, start-up vibration, motor current during load changing, shutdown voltage as illustrated in their works in [6, 48, 56–58].

WT has also some drawbacks, e.g., selection of mother wavelet is quite arbitrary—this may introduce error in the detection parameters [59]. For lower order wavelet overlapping between bands and frequency response will be very poor. Some parts of fundamental frequency leaked into adjacent frequency bands to mask the lower side harmonics, [60]. Further in some cases, the edge distortion from the transform, make the detection of the lower frequency band (i.e., below supply frequency) difficult, especially when the starting transient is very fast, [61]. To overcome these constraints a

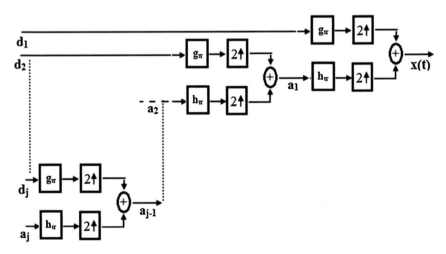

**Fig. 3.3** Multi-resolution reconstruction of discrete wavelet transforms

new methodology using envelope analysis of the signal through Hilbert transform has been proposed.


### 3.3.4  Hilbert Transform (HT)

The Hilbert transform is defined as

$$H[s(t)] = \frac{1}{\pi} \int_{-\infty}^{\infty} \frac{s(\tau)}{(t-\tau)} d\tau \tag{3.6}$$

Using the mean value theorem, we can evaluate (3.6) as follows:

$$H[s(t)] = \frac{1}{\pi t} \otimes s(t) \tag{3.7}$$

Thus Hilbert transform, $H[s(t)]$, of a signal $s(t)$ is obtained by the convolution of the function $(1/\pi t)$ and the original signal $s(t)$. $H[s(t)]$ and $s(t)$ are supposed to be in quadrature because theoretically they are out of phase by $\pi/2$. Also Hilbert transform is equivalent to the positive frequencies from the spectrum of $s(t)$ shifted by $-\pi/2$ and their amplitudes are doubled and the negative frequencies are removed. The Hilbert transform can be viewed as a filter which has the property to eliminate the negative frequencies and retain the positive frequencies with their phase shift of $\pi/2$.

Hilbert transform is used to obtain analytic signal from a real signal, for envelope analysis of transient current, etc. Motor current at steady state has also been analyzed by this method. Hilbert transform has been applied in [68–71].

*Analytic signal*: The complex form of a signal is known as analytic signal. The unique complex representation of a real signal, $s(t)$ is given by

$$z(t) = s(t) + j[H[s(t)]] \tag{3.8}$$

If the signal is of the form of $\alpha(t) \cdot \cos \Phi(t)$ like a real frequency modulated signal then its complex analytical signal, $z(t)$ may be given by

$$z(t) = \alpha(t)e^{j\Phi(t)} \tag{3.9}$$

*Envelope analysis*: The envelope of a complex signal, $z(t)$ is defined as

$$E(t) = |s(t) + jH[s(t)]|$$
$$\text{or } E(t) = \alpha(t) \tag{3.10}$$

The envelope signal occupies the low-frequency spectral region, the analysis of which gives better detection than the spectrum analysis of the original signal as the power frequency is eliminated from the signal.

### 3.3.5  CMS Rule Set [104]

#### 3.3.5.1  CMS Rule Set for Unbalance Assessment

If any two phase voltages and any two line currents of normalized data are plotted in voltage–voltage plane and current–current plane respectively, elliptical patterns are formed in these planes. One such pattern formed by voltage signals has been shown in Fig. 3.4. The unbalance in a system changes amplitudes and phase angles of the signals. This results in change in the length of major and minor axes of the closed patters.

The patterns in voltage–voltage and current–current plane carry the information of unbalance in their shape. To extract the features from the patterns, i.e., to get back the information of voltage or current unbalance from the patterns, the following parameters are introduced:

$$\text{at } x = X_{\text{MIN}} \text{ (point A)}, \quad y = Y_1 \text{ (AE)}$$
$$\text{at } x = X_{\text{MAX}} \text{ (point C)}, \quad y = Y_2 \text{ (CH)}$$

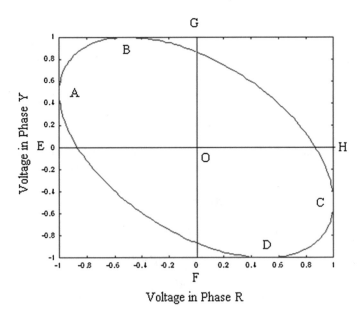

**Fig. 3.4**  Elliptical pattern in voltage–voltage plane

$$Y = Y_1 \sim Y_2 \tag{3.11}$$

at $y = Y_{\text{MIN}}$ (point D), $x = X_1$ (FD)
at $y = Y_{\text{MAX}}$ (point B), $x = X_2$ (GB)

$$X = X_1 \sim X_2 \tag{3.12}$$

For assessment of voltage unbalance in three-phase system, any two phase voltages are considered and plotted in voltage–voltage plane. For phase R and phase Y in voltage–voltage plane, phase R voltage is along $x$-axis and phase Y voltage is along $y$-axis. $X$ and $Y$ are calculated. They are denoted as $X_{\text{RY}}$ and $Y_{\text{RY}}$.

Similarly, for phase Y and B and phase B and R as $X_{\text{YB}}$ and $Y_{\text{YB}}$ and $X_{\text{BR}}$ and $Y_{\text{BR}}$ are calculated from respective loops in voltage–voltage plane. Thus, in three-phase system, two column matrices are obtained in voltage–voltage planes as follows:

$$[x_V] = \begin{bmatrix} X_{V_{\text{RY}}} \\ X_{V_{\text{YB}}} \\ X_{V_{\text{BR}}} \end{bmatrix} \tag{3.13}$$

$$[y_V] = \begin{bmatrix} Y_{V_{\text{RY}}} \\ Y_{V_{\text{YB}}} \\ Y_{V_{\text{BR}}} \end{bmatrix} \tag{3.14}$$

Note that these two matrices carry the features of pattern developed by two signals. Unbalance changes the dimensions of elliptical pattern. Changes in the length of major axis and minor axis of the elliptical patterns are reflected in the values of $[x_V]$ and $[y_V]$.

Similarly, for assessment of current unbalance in three-phase system, any two line currents are considered and plotted in current–current plane. For phase R and phase Y in current–current plane, phase R current is considered along $x$-axis and phase Y current is considered along $y$-axis. $X$ and $Y$ are calculated as mentioned above. In same way, $X$ and $Y$ are calculated with Y and B phase currents and B and R phase current in current–current planes. Two matrices are formed for current unbalance assessment as follows:

$$[x_I] = \begin{bmatrix} X_{I_{\text{RY}}} \\ X_{I_{\text{YB}}} \\ X_{I_{\text{BR}}} \end{bmatrix} \tag{3.15}$$

$$[y_I] = \begin{bmatrix} Y_{I_{\text{RY}}} \\ Y_{I_{\text{YB}}} \\ Y_{I_{\text{BR}}} \end{bmatrix} \tag{3.16}$$

Current unbalance changes the dimensions of elliptical pattern. Changes in the length of major axis and minor axis of the elliptical patterns are reflected in the values of $[x_V]$ and $[y_V]$. These two matrices are used for unbalance assessment.

CMS rule set for unbalance assessment using these matrices has been presented in Table 3.1.

**Table 3.1** CMS rule set for unbalance assessment

| Rule nos. | Rule | Inference |
|---|---|---|
| Rule 1 | $[x_V] = [1\ 1\ 1]$ and $[y_V] = [1\ 1\ 1]$ | Power system is perfect balanced |
| Rule 2 | $[x_V] \neq [1\ 1\ 1]$ and $[y_V] \neq [1\ 1\ 1]$ | Power system is unbalanced |
| Rule 3 | If Rule 2 is true and $X_{RY} \neq 1$ and $Y_{BR} \neq 1$ | Unbalance has occurred in phase R and percentage of unbalance $= (1 - X_{RY}) \times 100\ \%$ |
| Rule 4 | If Rule 2 is true and $X_{YB} \neq 1$ and $Y_{RY} \neq 1$ | Unbalance has occurred in phase Y and percentage of unbalance $= (1 - X_{YB}) \times 100\ \%$ |
| Rule 5 | If Rule 2 is true and $X_{BR} \neq 1$ and $Y_{YB} \neq 1$ | Unbalance has occurred in phase B and percentage of unbalance $= (1 - X_{BR}) \times 100\ \%$ |
| Rule 6 | If Rule 2 is true and $X_{RY} \neq 1, X_{YB} \neq 1$ $Y_{RY} \neq 1$ and $Y_{BR} \neq 1$ | Unbalance has occurred in phases R and Y, the percentage of unbalance in R phase $= (1 - X_{RY}) \times 100\ \%$ and the percentage of unbalance in Y phase $= (1 - X_{YB}) \times 100\ \%$ |
| Rule 7 | If Rule 2 is true and $X_{YB} \neq 1$, $X_{BR} \neq 1, Y_{YB} \neq 1$ and $Y_{RY} \neq 1$ | Unbalance has occurred in phases Y and B, the percentage of unbalance in Y phase $= (1 - X_{YB}) \times 100\ \%$ and the percentage of unbalance in B phase $= (1 - X_{BR}) \times 100\ \%$ |
| Rule 8 | If Rule 2 is true and $X_{BR} \neq 1$, $X_{RY} \neq 1, Y_{YB} \neq 1, Y_{BR} \neq 1$ | Unbalance has occurred in phases B and R, the percentage of unbalance in B phase $= (1 - X_{BR}) \times 100\ \%$ and the percentage of unbalance in R phase $= (1 - X_{RY}) \times 100\ \%$ |
| Rule 9 | If Rule 2 is true and $X_{RY} \neq 1$, $X_{YB} \neq 1, X_{BR} \neq 1, Y_{RY} \neq 1, Y_{YB} \neq 1$ and $Y_{BR} \neq 1$ | Unbalance has occurred in phases R, Y, and B. the percentage of unbalance in R phase $= (1 - X_{RY}) \times 100\ \%$, the percentage of unbalance in Y phase $= (1 - X_{YB}) \times 100\ \%$ and the percentage of unbalance in B phase $= (1 - X_{BR})\ 100\ \%$ |

### 3.3.5.2  CMS Rule for Determination of Highest Order Dominating Harmonics

Cleavages appear in the pattern formed by phase voltage in voltage–voltage plane and by line currents in current–current planes. The number of cleavages is used in **CMS Rule for Determination of Highest order Dominating Harmonics** as presented in Table 3.2.

### 3.3.5.3  CMS Rule Set for Harmonic Assessment in Clarke and Park Plane

By assessing the angular position of the cleavages appeared in the pattern formed Clarke and Park planes harmonics present in the signal are analyzed using **CMS Rule set for harmonic assessment in Clarke and Park plane** as presented in Table 3.3.

### 3.3.5.4  CMS Equations for Total Harmonic Distortion Factors

Harmonic distortions in Park plane are measured at steady-state condition by CMS equations of total harmonic distortions in Park Plane as given below:

**Table 3.2** CMS rule for determination of highest order dominating harmonics

| $C$ = number of cleavages in a pattern | Order of highest harmonic ($n_H = C + 1$) |
|---|---|

**Table 3.3** CMS rule set for harmonic assessment in Clarke and Park plane

| Rule no | Rule |
|---|---|
| 1 | If cleavage appears, then there is harmonic in the system |
| 2 | If rule 1 is true and if the number of cleavages is $C$ and order of harmonic is $n$, then $n = C + 1$ |
| 3 | If rule 1 is true, then there will be at least one cleavage at an angle $\theta_C$ given by $\alpha_n = 270° + \theta_C/2$ for even order ($n$) or, $\alpha_n = 270° + \theta_C$ for odd order ($n$) where, $\theta_C = 360°/C$ |
| 4 | If rules 1, 2 and 3 are true, then percentage amplitude of harmonic ($P$) is proportional to depth of locus along angle ($\alpha_n$) |
| 5 | Phase angle difference of the harmonic component from the fundamental is equal to the shift of the cleavage's angle multiplied by the number of cleavages found in Clarke plane, i.e., $\varphi_n = C \times \varphi_c$ where $C = N - 1$ |

Total harmonic distortion factor for voltage along d axis

$$
\mathrm{THD}_{V_d} = \frac{\sqrt{\sum_{m=2,3,\dots}(V_{dm})^2}}{V_{d1}}
$$

$$
= \frac{\sqrt{\sum_{m=2,3,\dots}\left[\left\{K_1 A_{dm}^{(v_d-v_{REF})}\right\}^2 + \left\{K_2 A_{dm}^{(v_{REF}\,v_d\,-t)}\right\}^2\right]}}{\sqrt{\left[\left\{K_1 A_{d1}^{(v_d-v_{REF})}\right\}^2 + \left\{K_2 A_{d1}^{(v_{REF}\,v_d\,-t)}\right\}^2\right]}}
\tag{3.17}
$$

Total harmonic distortion factor for current along $d$-axis is written as

$$
\mathrm{THD}_{I_d} = \frac{\sqrt{\sum_{n=2,3,\dots}(I_{dn})^2}}{I_{d1}}
$$

$$
= \frac{\sqrt{\sum_{n=2,3,\dots}\left[\left\{K_1 A_{dn}^{(i_d-i_{REF})}\right\}^2 + \left\{K_2 A_{dn}^{(i_{REF}i_d-t)}\right\}^2\right]}}{\sqrt{\left[\left\{K_1 A_{d1}^{(i_d-i_{REF})}\right\}^2 + \left\{K_2 A_{d1}^{(i_{REF}i_d-t)}\right\}^2\right]}}
\tag{3.18}
$$

Total harmonic distortion factor for voltage along $q$-axis

$$
\mathrm{THD}_{V_q} = \frac{\sqrt{\sum_{m=2,3,\dots}(V_{qm})^2}}{V_{q1}}
$$

$$
= \frac{\sqrt{\sum_{m=2,3,\dots}\left[\left\{K_1 A_{qm}^{(v_q-v_{REF})}\right\}^2 + \left\{K_2 A_{qm}^{(v_{REF}v_q-t)}\right\}^2\right]}}{\sqrt{\left[\left\{K_1 A_{q1}^{(v_q-v_{REF})}\right\}^2 + \left\{K_2 A_{q1}^{(v_{REF}v_q-t)}\right\}^2\right]}}
\tag{3.19}
$$

Total harmonic distortion factor for current along $q$ axis

$$
\mathrm{THD}_{I_q} = \frac{\sqrt{\sum_{n=2,3,\dots}(I_{qn})^2}}{I_{q1}}
$$

$$
= \frac{\sqrt{\sum_{n=2,3,\dots}\left[\left\{K_1 A_{qn}^{(i_q-i_{REF})}\right\}^2 + \left\{K_2 A_{qn}^{(i_{REF}\,i_q-t)}\right\}^2\right]}}{\sqrt{\left[\left\{K_1 A_{q1}^{(i_q-i_{REF})}\right\}^2 + \left\{K_2 A_{q1}^{(i_{REF}\,i_q-t)}\right\}^2\right]}}
\tag{3.20}
$$

In the above equations, '$K$'s are constant and '$A$'s are areas formed in different planes in Park domain.


### 3.3.6 Radar Analysis of Stator Current Concordia

Radar analysis of stator current Concordia is another method for diagnosis of faults in squirrel cage induction motor. A number of research works [72–76] have been done by the use of this simple method. Diagnosis of current unbalance at the terminals of a three-phase induction motor is done in [72] by assessing feature pattern extraction method by S. Chattopadhyay et al. In [73] current Concordia pattern-based Fuzzy decision system has been used by Fatiha Zidani et al. to diagnose stator fault of an induction motor. Fault detection and diagnosis in an induction machine have also been done in Park plane using Concordia of stator current vector [74, 75]. Here stator current is transformed into direct–quadrature ($d$–$q$) plane by Park matrix. In this plane, harmonics free stator current Concordia is an exact circular pattern. But the presences of harmonics bring cleavages in this Concordia. Harmonics assessment using Concordia in Park plane is done by Chattopadhyay et al. in [76] where CMS rule set has been suggested for identifying the presence of harmonic in the current which states that if cleavage appears in the Concordia then there is harmonic and the order of the harmonic is one more than the number of cleavages, i.e.,

$$n = C + 1 \qquad (3.21)$$

where $n$ is harmonic and $C$ is number of cleavages present in the Concordia.


## 3.4 Research Trend in Fault Analysis

In Sect. 3.2.3 different faults those may happen in an induction motor have been reviewed. Different fault analysis techniques and signal processing tools are reviewed in Sects. 3.3.1 and 3.3.2 respectively. In this section, trend of research works by researchers, in the analysis of different faults using different signal processing tools in the last few years, have been reviewed.

***Broken rotor-bar fault***: W. Deleroi has analyzed the fault, in his work [77], by superimposing the fault currents of the concerned motor. In [45] da Silva et al. have done envelope analysis of the stator current to diagnose both broken rotor bar fault and stator short-circuit fault. Sian et al. in [46] have done spectral analysis of stator current and flux for detection of broken rotor bar fault. In [52] Supangat et al. have shown the method of detection of this fault by analyzing the stator current and the effects of loading. Mirafzal et al. have shown, in their work [78], the diagnosis of this fault by the use of the rotor magnetic field space vector orientation. Ouma et al., in their work [68], have analyzed the neutral voltage of the induction motor to diagnose the fault. In [69], Lebaroud et al. have shown an improved technique of detection of the fault. They have taken the signal after disconnecting the motor supply for the analysis purpose. Arabaci et al. have used motor power factor to diagnose the fault in [80]. Nemec et al. in their work [81], have shown the detection of fault through the analysis of voltage modulation.

*Air gap eccentricity fault*: Dorrell et al. have analyzed, in their work [82], air gap flux, current and vibration signals of an induction motor for the detection of static and dynamic eccentricity of the motor. In [83], Drif et al. have diagnosed the air gap eccentricity fault by the analysis of complex apparent power signature. Ahmed et al. have proposed in [84] the detection technique of the fault by the analysis of frequency spectrum. In [85], Ozelgin has analyzed magnetic flux density for diagnosis of air gap eccentricity fault and bearing fault. Thomson et al. have performed online current monitoring in their work [86] and have proposed the method of prediction of the level of air gap eccentricity by the application of Finite Element. Using stator current Wigner distributions, Blodt et al. in [87], have distinguished the load torque oscillations from eccentricity faults in induction motor. In [88], Faiz et al. have analyzed mixed eccentricity fault using Finite Element. In [89], Nandi et al. have done performance analysis of a three-phase induction motor suffering from mixed eccentricity fault.

*Stator winding fault*: Lee et al. have described, in their work [90], a method for detection of stator turn fault by monitoring the sequence component impedance matrix. Cash et al. have described the method to predict insulation failure in the motor using line-to-neutral voltage in [16]. In [45], da Silva et al. have shown a method for stator short-circuit fault diagnostics based on three-phase stator current envelopes. A current Concordia pattern-based Fuzzy decision system has been described in [73] by Zidani et al. For detection of the stator inter-turn fault, Benbouzid et al. have worked using artificial neural network in [91], Mirafzal et al. have worked using Pendulous Oscillation phenomenon in [92], Ballal et al. have used adaptive neural Fuzzy Interference system in [93], Stone et al. have used in [94] partial discharge test. Ukil et al., in their work [95], have used motor current zero crossing instants to detect the fault. Here analysis using zero crossing time (ZCT) signal of stator current is presented. Negative sequence currents are also used to detect stator winding fault [96, 97]. In [97], Kato et al. have shown the method of diagnosis of stator winding turn fault by direct detection of negative sequence currents of the induction motor. Protection method for induction motor against faults due to voltage unbalance and single phasing has been described by Sudha et al. in [98].

*Wavelet Transform*: Using this tool of data signal processing a number of research works have been performed so far for MCSA technique. In this technique mainly transient signature is analyzed. In [48, 55, 59–61], WT has been used to analyze transient current of motor. Douglas et al. have shown in [59] the importance of selection of wavelet to analyze the current signature. Broken rotor bar fault has been diagnosed in [55, 60, 61] applying WT. In [49] Lu et al. have performed wavelet analysis of one-cycle average power to diagnose rotor fault. Ye et al. in [62] have analyzed current signature using wavelet packet decomposition for studying mechanical faults of induction motor. Cusido et al. in their works [57, 58], have used power spectral density after performing the WT to detect motor faults. In [99] Mohammed et al. have used Finite Element and DWT simultaneously for modeling and characterization of internal faults of motor. In [100] Pons-Llinares et al. have used analytic WT of transient currents to diagnose eccentricity fault of inverter-fed induction motor.

***Hilbert Transform***: This eliminates some of the limitations of the WT as discussed in Sects. 3.2.3 and 3.2.4. For MCSA technique this data signal processing tool has been used by a number of researchers. Puche-Panadero has used this tool in his work [51] for MCSA to diagnose rotor asymmetries. For this purpose he collected the current signature when the motor was running at very low slip. Ouma et al. have worked using this tool to detect broken bar fault in [68]. They have used neutral voltage of the motor as signal. Motor fault detection has been performed using Hilbert transform by Medoued et al. in [69] and by Jimenez et al. in [71]. Jimenez et al. have used both WT and Hilbert Transform tools simultaneously for signal analysis. Pons-Llinares et al. in [70] for the purpose of diagnosing eccentricity fault have analyzed motor starting current using this Hilbert transform tool.

***Park vector pattern recognition***: Normally three-phase (R–Y–B) currents are available from an induction motor. To obtain direct and quadrature axes currents, these three-phase currents are transformed into $d$–$q$ plane using Park's matrix as given in (3.12)

$$\begin{pmatrix} i_d \\ i_q \end{pmatrix} = \sqrt{\frac{2}{3}} \times \begin{pmatrix} \cos\theta & \cos(\theta - {}^{2\pi}\!/_3) & \cos(\theta - {}^{4\pi}\!/_3) \\ -\sin\theta & -\sin(\theta - {}^{2\pi}\!/_3) & -\sin(\theta - {}^{4\pi}\!/_3) \end{pmatrix} \times \begin{pmatrix} i_a \\ i_b \\ i_c \end{pmatrix}$$

$$(3.22)$$

Some researchers have worked out [74, 75, 101–103] with these direct axis ($i_d$) and quadrature axis ($i_q$) currents and analyzing the pattern formed by these currents. In [74] Nejjari et al. have used current Park's vector pattern to monitor and diagnose electrical faults of induction motors. Diallo et al. have shown, in their work [75], a pattern recognition approach based on Concordia of stator mean current vector. Cardoso et al. have used Park's vector approach, in [101] to predict the level of air gap eccentricity in a three-phase induction motor and in [102] to diagnose inter-turn stator winding fault. In [103] Szabo et al. have shown the detection method of rotor fault of a squirrel cage induction machine by Park's vector approach. CMS rule set is found as an effective tool for harmonic assessment in Park plane [104].

## 3.5  Conclusion

From literature survey it is observed that for fault detection of induction motor, techniques like MCSA, vibration analysis, thermal analysis, etc., are used. Out of these, in maximum cases MCSA is used due to its different advantages. In this book also, MCSA will be used for detection of fault in induction motor.

Detailed review reveals that for MCSA, any of the motor currents, namely starting current, steady-state current, or shutdown current may be used as signal for analysis. Each of the currents as signal has some advantages or disadvantages over the other as discussed in this chapter. In this work both the starting current and steady-state current of the motor under load and no-load conditions will be used as signal for MCSA.

Researchers have shown that various signal processing tools can be used successfully under different running conditions such as 'starting' or 'steady' at 'under load' or 'no-load' of the motor for analysis of the current signal in this MCSA technique. Each tool has some advantages or disadvantages in a particular condition of the motor. In this book, FFT, WT, Hilbert Transform, Radar analysis of current Concordia, and feature pattern extraction method will be used on stator current signature of induction motor.

Different faults have been studied by researchers as seen in the literature review. This work is confined in the diagnoses of (i) broken rotor bar fault, (ii) rotor mass unbalance fault, (iii) stator winding fault, (iv) single-phasing fault, and (v) one manufacturing defect called crawling of an induction motor. At the end of this book, a comparative study has been presented showing the results of all the diagnosis techniques for each of the five faults mentioned above. Attempts have also been made to choose the best option for the above.

# References

1. Rajagopal MS, Setharaman KN, Ashwathnarayana PA (1998) Transient thermal analysis of induction motors. IEEE Trans Energy Convers 13(1):62–69. ISSN: 0885-8969
2. Cezario CA, Verardi M, Borges SS, da Silva JC, Oliveira AAM (2005) Transient thermal analysis of an induction electric motor. In: 18th international congress of mechanical engineering. OuroPreto, MG, 6–11 Nov 2005
3. Mellor PH, Roberts D, Turner DR (1991) Lumped parameter thermal model for electrical machines of TEFC design. IEEE Proc Electr Power Appl 138:205–218
4. Boys JT, Miles MJ (1994) Empirical thermal model for inverter-driven cage induction machines. Inst Electr Eng Proc Electr Power Appl 141(6):360–372
5. Zhang P, Du Y, Habetler TG, Lu B (2011) Magnetic effects of DC signal injection on induction motors for thermal evaluation of stator windings. IEEE Trans Ind Electron 58(5)
6. Cusido J, Roura I, Martinez JLR (2014) Transient analysis and motor fault detection using the wavelet transform. www.intechopen.com
7. Kapoor SR (2013) Commonly occurring faults in three phase induction motors—causes, effects and detection—a review. J Inf Knowl Res Electr Eng 2(2):178–185. ISSN 0975-6736
8. Ellison AJ, Young SJ (1971) Effects of rotor eccentricity on acoustic noise from induction machines. Proc IEE 118(1):174–184
9. Lee YS, Nelson JK, Scarton HA, Teng D, Ghannad SA (1994) An acoustic diagnostic technique for use with electric machine insulation. IEEE Trans Dielectr Electr Insul 1(6):1186–1193
10. Oskouel ET, Roddis AJ (2007) A condition monitoring device using acoustic emission sensors and data storage devices. UK Patent Application GB 2340034 A
11. da Silva AM, Povinelli RJ, Demerdash NAO (2013) Rotor bar fault monitoring method based on analysis of air-gap torques of induction motor. IEEE Trans Industr Inf 9(4):2274–2283

12. Stopa MM, de Jusus Cardoso Filho B (2012) Load torque signature analysis: an alternative to MCSA to detect faults in motor driven loads. IEEE Energy Conversion Congress and Exposition (ECCE), 15–20 Sept 2012, pp 4029–4036
13. Hsu JS (1995) Monitoring of defects in induction motors through air-gap torque observation. IEEE Trans Ind Appl 31(5):1016–1021
14. Arabaci H, Bilgin O (2010) Effects of rotor faults in squirrel cage induction motor on the torque-speed curve. In: XIX international conference on electrical machines (ICEM), Rome, 6–8 Sept 2010, ISBN (print): 978-1-4244-4174-7, pp 1–5
15. Tavner PJ, Gaydon BG, Ward DM (1986) Monitoring generators and large motors. Proc Inst Elect Eng B 133(3):169–180
16. Cash MA, Habetler HG, Kliman GB (1998) Insulation failure prediction in AC machines using line-neutral voltages. IEEE Trans Ind Appl 34(6):1234–1239
17. Bonnett AH, Soukup GC (1992) Cause and analysis of stator and rotor failures in three phase squirrel-cage induction motors. IEEE Trans Ind Appl 28(4):921–937
18. Edwards DG (1994) Planned maintenance of high voltage rotating machine insulation based upon information derived from on-line discharge measurements. Proc IEE Int Conf Life Manage Power Plants 401:101–107
19. Stone GC, Floyd BA, Campbell SR, Shedding HG (1997) Development of automatic, continuous partial discharge monitoring systems to detect motor and generator partial discharges. Proceedings IEEE, IEMDC'97, pp MA2-3.1–MA2-3.3
20. Stone GC, Sedding HG, Costello MJ (1996) Application of partial discharge testing to motor and generator stator winding maintenance. IEEE Trans Ind Appl 32(2):459–464
21. Tetrault SM, Stone GC, Sedding HG (1999) Monitoring partial discharges on 4KV motor windings. IEEE Trans Ind Appl 35(3):682–688
22. Belahcen A, Arkkio A, Klinge P, Linjama J, Voutilainen V, Westerlund J (1999) Radial forces calculation in a synchronous generator for noise analysis. In: Proceedings of the third Chinese international conference on electrical machines, Xi'an, China, August 1999, pp 119–122
23. Mehala N (2010) Condition monitoring and fault diagnosis of induction motor using motor current signature analysis. A PhD thesis submitted to the Electrical Engineering Department, NIT Kurukshetra, India, Oct 2010
24. Iorgulescu M, Beloiu R (2008) Vibration and current monitoring for faults diagnosis of induction motors. Annals of the University of Craiova, electrical engineering series, No. 32, ISSN 1842-4805
25. Su H, Chong KT, Kumar RR (2011) Vibration signal analysis for electrical fault detection of Induction machine using neural networks. Neural Comput Appl 20(2):183–194
26. Widodo A, Yang B-S, Han T (2007) Combination of independent component analysis and support vector machines for intelligent fault diagnosis of induction motor. Expert Syst Appl 32(2):299–312
27. Maruthi GS, Hegde V (2013) An experimental investigation on broken rotor bar in three phase induction motor by vibration signature analysis using MEMS accelerometer. Int J Emerg Technol Adv Eng 3(4):357–363. ISSN 2250-2459
28. da Costa C, Mathias MH, Ramos P, Girao PS (2010) A new approach for real time fault diagnosis in induction motors on vibration measurement. 978-1-4244-2833-5/10/ (C), IEEE
29. Wang C, Lai JCS (1999) Vibration analysis of an induction motor. J Sound Vib 224(4):733–756
30. McInerny SA, Dai Y (2003) Basic vibration signal processing for bearing fault detection. IEEE Trans Educ 46(1):149–156
31. Tsypkin M (2011) Induction motor condition monitoring: vibration analysis technique—a practical implementation. In: Proceedings IEEE international conference on electric machine and drives conference (IEMDC), USA, 15–18 May 2011. ISBN 978-1-4577-0060-6
32. Li W, Mechefske CK (2004) Induction motor fault detection using vibration and Stator current methods. Insight 46(8):473–484
33. Onel IY, Dalci KB, Senol I (2005) Detection of outer raceway bearing defects in small induction motor using stator current analysis. Sadhana 30:713–722, Part 6

34. Bellini A, Concari C, Franceschini G, Lorenzani E, Tassoni C, Toscani A (2006) Thorough understanding and experimental validation of current sideband components in induction machines rotor monitoring. In: IECON 2006–32nd annual conference on IEEE industrial electronics, pp 4957–4962
35. Watson JF, Paterson NC (1998) Improved techniques for rotor fault detection in three phase induction motors. 0-7803-4943-1/98 ©1998, IEEE
36. Kliman GB, Stein J (1992) Methods of motor current signature analysis. Electric Mach Power Syst 20(5):463–474
37. King GJ (2010) Induction motor fault detection using the fast orthogonal search algorithm. A thesis submitted for the Degree of Master of Applied Science in Electrical Engineering, Royal Military College of Canada, April
38. Filippetti F, Martelli M, Franceschini G, Tassoni C (1992) Development of expert system knowledge base to on-line diagnosis of rotor electrical faults of induction motors. Industry Applications Society Annual Meeting, 1992. Conference Record of the 1992 IEEE, pp 92–99
39. Benbouzid MEH (2000) A review of induction motor signature analysis as a medium for fault detection. IEEE Trans Industr Electron 47(5):984–993
40. Thomson WT, Gilmore RJ (2003) Motor current signature analysis to detect faults in induction motor drives-fundamentals, data interpretation, and industrial case histories. In: Proceedings of the thirty-second turbomachinery symposium, pp 145–156
41. Jung J-H, Lee J-J, Kwon B-H (2006) Online diagnosis of induction motors using MCSA. IEEE Trans Industr Electron 53(6):1842–1852
42. Ahmed I, Supangat R, Grieger J, Ertugrul N, Soong WL (2004) A baseline study for online condition monitoring of induction machines. Australian universities power engineering conference (AUPEC), Brisbane, Australia, 2004
43. Thomson WT, Rankin D, Dorrell DG (1999) On-line current monitoring to diagnose air gap Eccentricity in large three-phase induction motors—industrial case histories to verify the predictions. IEEE Trans Energy Convers 14(4):1372–1378
44. Thomson WT, Barbour A (1998) On-line current monitoring and application of a finite element method to predict the level of air gap eccentricity in 3-phase induction motor. IEEE Trans Energy Convers 13(4):347–357
45. da Silva AM, Povinelli RJ, Demerdash NAO (2006) Induction machine broken bar and stator short-circuit fault diagnostics based on three phase stator current envelopes. In: IEEE transactions on industrial electronics
46. Sian J, Graff A, Soong WL, Ertugrul N (2003) Broken bar detection in induction motors using current and flux spectral analysis. In: Australian universities power engineering conference (AUPEC), Christchurch, New Zealand
47. Bellini A, Filippetti F, Franceschini G, Tassoni C, Kliman GB (2001) Quantitative evaluation of induction motor broken bars by means of electrical signature analysis. IEEE Trans Ind Appl 37:1248–1255
48. Douglas H, Pillay P, Ziarani AK (2004) A new algorithm for transient motor current signature analysis using wavelets. IEEE Trans Industry Appl 40(5):1361–1368
49. Lu B, Paghda M (2008) Induction motor rotor fault diagnosis using wavelet analysis of one-cycle average power. 978-1-4244-1874-9/08 ©2008, IEEE
50. Bonnet AH, Soukup GC (1986) Rotor failures in squirrel cage induction motors. IEEE Trans Ind Appl IA-22(6):1165–1173
51. Puche-Panadero R (2009) Improved resolution of the MCSA method via Hilbert transform, enabling the diagnosis of rotor asymmetries at very low slip. IEEE Trans Energy Convers 24(1):52–59
52. Supangat R, Ertugrul N, Soong WL, Gray DA, Hansen C, Griegar J (2006) Detection of broken rotor bars in induction motor using starting current analysis and effects of loading. In: IEE Proceedings of electrical power application, vol 153(6), November 2006
53. Gaeid KS, Mohamed HAF (2010) Diagnosis and fault tolerant control of the induction motors techniques a review. Australian J Basic Appl Sci 4(2):227–246. ISSN 1991-8178

54. Gandhi A, Corrigan T, Parsa L (2011) Recent advances and modeling and online detection of stator inter turn faults in electrical motors. IEEE Trans Industr Electron 58(5):1564–1573
55. Mehala N, Dahiya R (2009) Rotor faults detection in induction motor by wavelet analysis. Int J Eng Sci Technol 1:90–99
56. Antonino JA, Riera M, Felch JR, Climente V Study of the startup transient for the diagnosis of broken bars in induction motors: a review. http://www.aedie.org/9CHLIE-paper-send/318_Antonino.pdf
57. Cusido J, Jornet A, Romeral L, Ortega JA, Garcia A (2006) Wavelet and PSD as a fault detection technique. ITMC 2006-instrumentation and measurement technology conference. Sorrento, Italy, 24–27 April 2006
58. Cusido J, Romeral L, Ortego JA, Rosero JA, Espinosa AG (2008) Fault detection in induction machines using power spectral density in wavelet decomposition. IEEE Trans Ind Electron 55(2)
59. Douglas H, Pillay P (2005) The impact of wavelet selection on transient motor current signature analysis. 0-7803-8987-5/05 ©2005, IEEE
60. Daviu JA, Guasp MR, Folch JR, Molina PM (2006) Validation of a new method for the diagnosis of rotor bar failures via wavelet transformation in industrial induction machines. IEEE Trans Ind Appl 42(4):990–996
61. Daviu JA,Guasp MR, Folch JR, Gimenez FM, Peris A (2006) Application and optimization of the discrete wavelet transform for the detection of broken rotor bars in induction machines. Appl Comput Harmon Anal 21(2):268–279
62. Ye Z, Wu B, Sadeghian A (2003) Current signature analysis of induction motor mechanical faults by wavelet packet decomposition. IEEE Trans Ind Electron 50(6)
63. Menacer A et al (2004) Stator current analysis of incipient fault into asynchronous motor rotor bars using Fourier fast transform. J Electr Eng 55(5–6):122–130
64. Schoen RR, Lin BK, Habetler TG, Schlag JH, Farag S (1995) An unsupervised on-line system for induction motor fault detection using stator current monitoring. IEEE Trans Ind Appl 31:1280–1286
65. Marcelo C, Fossatti JP, Terra JI (2012) Fault diagnosis of induction motors based on FFT. www.intechopen.com
66. Cusido J, Rosero J, Aldabas E, Ortega JA, Romeral L (2006) New fault detection techniques for induction motors. Electr Power Qual Utilisation, Mag II(1)
67. Da Silva AA et al (1997) Rotating machinery monitoring and diagnosis using short time Fourier transform and wavelet techniques. In: Proceedings of international conference maintenance and reliability, Knoxville, TN, vol 1, pp 14.01–14.15
68. Ouma ME, Khezzar AA, Boucherma M, Razik H, Andriamalala RN, Baghh L (2007) Neutral voltage analysis for Broken-bar detection in Induction motors using Hilbert transform phase, IEEE, pp 1940–1947
69. Medoued A, Lebaroud A, Sayad D (2013) Application of Hilbert transform to fault detection in electric machines. Adv Differ Eqn 2013:2 doi:10.1186/1687-1847-2013-2
70. Pons-Llinares J, Roger-Folch J, Pineda-Sanchez M (2009) Diagnosis of eccentricity based on the Hilbert transform of the startup transient current. In: IEEE international symposium on diagnostics on electrical machines, power electronics and drives, SDEMPED 2009
71. Jimenez GA et al (2007) Fault detection in induction motors using Hilbert and Wavelet transforms. Electr Eng 89:205–220. doi:10.1007/s00202-005-0339-6
72. Chattopadhyay S, Nandi S, Sengupta S (2004) A novel approach for diagnosis of current unbalance at the terminals of a three-phase induction motor. In: Proceedings of MS2004-international conference on modeling and simulation by AMSE, Lyon France, 5–7 July 2004, pp 3.22–3.25
73. Zidani F, Benbouzid MEH, Diallo D, Naït-Saïd MS (2003) Induction motor stator faults diagnosis by a current Concordia pattern-based fuzzy decision system. IEEE Trans Energy Convers 18(4):469–475

74. Nejjari H, Benbouzid MEH (2000) Monitoring and diagnosis of induction motors electrical faults using a current Park's vector pattern learning approach. IEEE Trans Ind Appl 36(3):730–735
75. Diallo D, Benbouzid MEH, Hamad D, Pierre X (2005) Fault detection and diagnosis in an induction machine drive: a pattern recognition approach based on Concordia stator mean current vector. IEEE Trans Energy Convers 20(3):512–519
76. Chattopadhyay S, Mitra M, Sengupta S (2007) Harmonic analysis in a three-phase system using Park transformation technique. AMSE Int J Model Simul 80(3):42–58
77. Deleroi W (1984) Broken bars in squirrel cage rotor of an induction motor-part I: description by superimposed fault currents. Arch Elektrotech 67:91–99
78. Mirafzal B, Demerdash NAO (2004) Induction machine broken-bar fault diagnosis using the rotor magnetic field space-vector orientation. IEEE Trans Ind Appl 40:534–542
79. Lebaroud A, Bentounsi A (2005) Detection improvement of the broken rotor bars of induction motor after supply disconnection. J Electr Eng 56(11-12):322–326
80. Arabacı H, Bilgin O, Ürkmez A (2011) Rotor bar fault diagnosis by using power factor. In: Proceedings of the world congress on engineering 2011, vol II, WCE, 6–8 July 2011, London, UK
81. Nemec M, Ambrozic V, Nedeljkovic D, Fiser R (2006) Detection of broken-bars in induction motor through the analysis of voltage modulation. IEEE ISIE, Montreal, Quebec, Canada, 9–12 July 2006, pp 2450–2454
82. Dorrell DG, Thomson WT, Roach S (1997) Analysis of air-gap flux, current and vibration signals as function of a combination of static and dynamic eccentricity in 3-phase induction motors. IEEE Trans Ind Appl 33:24–34
83. Drif M, Cardoso AJM (2008) Air gap eccentricity fault diagnosis in three phase induction motor by the complex apparent power signature analysis. IEEE Trans Ind Electron 55(3)
84. Ahmed I, Ahmed M, Imran K, Khan MS, Akhtar SJ (2011) Detection of eccentricity faults in machine using frequency spectrum technique. Int J Comput Electr Eng 3(1)
85. Ozelgin I (2008) Analysis of magnetic flux density for airgap eccentricity and bearing faults. Int J Syst Appl Eng Dev 2(4)
86. Thomson WT, Barbour A (1998) On-line current monitoring and application of a finite element method to predict the level of airgap eccentricity in 3-phase induction motor. IEEE Trans Energy Convers 13(4):347–357
87. Blodt M, Regnier J, Faucher J (2006) Distinguishing load torque oscillations and eccentricity faults in induction motors using stator current Wigner distributions. IEEE, pp 1549–1557
88. Faiz J, Ebrahimi BM, Akin B, Toliyat HA (2008) Finite-element transient analysis of induction motors under mixed eccentricity fault. IEEE Trans Magn 44(1)
89. Nandi S, Bharadwaj R, Toliyat HA, Parlos AG (1998) Performance analysis of a three phase induction motor under mixed eccentricity condition, IEEE, pp 123–128
90. Lee SB, Tallam RM, Habetler TG (2003) A robust on-line turn-fault detection technique for induction machines based on monitoring the sequence component impedance matrix. IEEE Trans Power Electron 18(3):865–872
91. Benbouzid MEH et al (1998) Induction motor inter turn short-circuit and bearing wear detection using artificial neural networks. Electromotion 5(1):15–20
92. Mirafzal B, Povinelli RJ, and Nabeel A.O. Demerdash, "Inter turn fault diagnosis in induction motors using the Pendulous Oscillation phenomenon. IEEE Trans Energy Convers 21(4):878–881
93. Ballal MS, Khan ZJ, Suryawansi HM, Sonolikar RL (2007) Adaptive neutral fuzzy interference system for the detection of inter-turn insulation and bearing wear faults in induction motor. IEEE Trans Ind Electron 54(1):250–259
94. Stone GC, Sedding HG (1995) In-service evaluation of motor and generator stator windings using partial discharge tests. IEEE Trans Ind Appl 31:299–303
95. Ukil A, Chen S, Andenn A (2011) Detection of stator short circuit faults in three phase induction motors using motor current zero crossing instants. Electr Power Syst Res 81(4):1036–1044

96. Oviedo S, Quiroga J, Borras C (2011) Motor current signature analysis and negative sequence current based stator winding short fault detection in an induction motor. Dyna, 170
97. Kato T, Inoue K, Yoshida K (2014) Diagnosis of stator-winding-turn faults of induction motor by direct detection of negative sequence currents. Electr Eng Jpn 186(3):75–84
98. Sudha M, Anbalagan P (2007) A novel protection method for induction motor against faults due to voltage unbalance and single phasing. In: 33rd annual conference of IEEE industrial electronics society (IECON), 5–8 Nov 2007, Taipai, Taiwan, pp 1144–1148
99. Mohammed OA, Abed NY, Ganu S (2006) Modeling and characterization of induction motor internal faults using finite element and discrete wavelet transform. IEEE Trans Magn 42(10)
100. Pons-Llinares J, Antonino-Daviu J, Roger-Folch J, Morinigo-Sotelo D, Duque-Perez O (2010) Eccentricity diagnosis in inverter–fed induction motors via the analytic wavelet transform of transient currents. In: XIX international conference on electrical machines (ICEM), Sept 2010
101. Cardoso AJM, Saraiva ES (1992) Predicting the level of air-gap eccentricity in operating three phase induction motor, by Park's vector approach. In: Proceedings of the IAS annual meeting, pp 132–135
102. Cardoso AJM, Cruz SMA, Fonseca DSB (1999) Inter-turn stator winding fault diagnosis in three-phase induction motors, by Park's vector approach. IEEE Trans Energy Convers 14(3):595–598
103. Szabo L, Kovacs E, Toth F, Fekete G (2007) Rotor faults detection methods for squirrel cage induction machines based on Park's vector approach. In: Ordea University annals, electro technical fascicle, computer science and control systems session, pp 234–239
104. Chattopadhyay S, Mitra M, Sengupta S (2011) Electric power quality, 1st edn. Springer, Berlin

# Chapter 4
# Broken Rotor Bar

**Abstract** The chapter deals with broken rotor bar fault. Diagnosis is done both using steady-state signal as well as starting current transients. Concordia is formed in Park plane using steady-state current and analyzed. The Concordia-based assessment is done in Park Plane using starting current transients. Then radars are formed in Park plane wherefrom the faults are differentiated compared with those of normal motor. Then harmonics generated due to rotor broken bar has been assessed using Hilbert and wavelet transforms. The main contribution in this chapter is the application of new fault diagnostic concept based on radar assessment and envelope analysis to extract low-frequency oscillation using Hilbert and wavelet transforms.

**Keywords** Broken rotor bar · Concordia · Discrete wavelet transform · FFT · Park plane · Radar

**Chapter Outcome**

After completion of the chapter, readers will be able to gather knowledge and information regarding the following areas:

- Broker rotor bar
- Steady-state approaches for broken rotor bar assessment
- Diagnosis of broken rotor bar using starting transients
- Concordia-based assessment
- Radar-based diagnosis
- Diagnosis using Hilbert and wavelet transforms.

## 4.1 Introduction

Induction motors are rugged and very efficient, yet they are subject to different types of undesirable faults. One of the most vulnerable parts for fault in induction motor is rotor bars. As per the statistical studies of IEEE and EPRI (Electric Power Research Institute), percentage of rotor fault occurrence possibility in induction motor is 8 and 9 %, respectively. Rotor faults in induction motors may be listed as

© Springer Science+Business Media Singapore 2016                                      57
S. Karmakar et al., *Induction Motor Fault Diagnosis*,
Power Systems, DOI 10.1007/978-981-10-0624-1_4

follows—(i) broken bar fault, (ii) rotor mass unbalance fault, and (iii) bowed rotor fault. Like any other faults in induction motor, rotor fault also results in unbalanced stator currents and voltages, oscillations in torque, reduction in efficiency and torque, overheating, and excessive vibration. Moreover, these motor faults can increase the magnitude of certain harmonic components of currents and voltages.

## 4.2   Broken Rotor Bar Fault

In an induction motor, broken rotor bar fault occurs when one or more of the rotor bars or end rings are partially cracked or completely broken. Causes of rotor broken bar fault may be one or more of the points, namely manufacturing defect, thermal stress and mechanical stress caused by bearing faults, frequent starts of the motor at rated voltage, and fatigue of metal of the rotor bar.

Cracked or broken bar fault produces a series of sideband frequencies [1, 2] in the stator current given by

$$f_{\text{brb}} = f(1 \pm 2ks) \qquad (4.1)$$

where $f$ is the supply frequency, $s$ is the slip, and $k$ is an integer.

Details about rotor broken bar have already been discussed in Chap. 2.

## 4.3   Diagnosis of Broken Rotor Bar Fault

For assessment of broken rotor bar fault, transient as well as steady-state stator current may be captured under load and no load conditions of the motor. These currents are analyzed using different signal processing tools in different methods, as shown in Fig. 4.1, for diagnosis of rotor broken bar fault of an induction motor. Methods, their experimentations, results, and observations are described in subsequent sections.

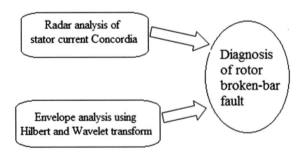

**Fig. 4.1** Diagnosis of rotor broken bar fault using different techniques

## 4.4 Diagnosis Through Radar Analysis of Stator Current Concordia [3]

Radar analysis of stator current Concordia used for diagnosis of cracks in rotor bar of a squirrel cage induction motor has been studied. For this purpose, stator current under starting condition is first stepped down through current transformer and then captured by a data acquisition system. Concordia of starting current is then developed in Park plane wherefrom primary diagnosis of faults is done. Then radar analysis is performed, wherefrom by comparison, motor with cracked rotor bar is isolated.

## 4.5 Concordia in Park Plane at Steady State

### 4.5.1 Concordia in Park Plane

Three phase stator currents can be transformed into direct–quadrature (d–q) plane by Park matrix as shown in (4.2):

$$\begin{pmatrix} i_d \\ i_q \end{pmatrix} = \sqrt{\frac{2}{3}} \times \begin{pmatrix} \cos\theta & \cos(\theta - 2\pi/3) & \cos(\theta - 4\pi/3) \\ -\sin\theta & -\sin(\theta - 2\pi/3) & -\sin(\theta - 4\pi/3) \end{pmatrix} \times \begin{pmatrix} i_a \\ i_b \\ i_c \end{pmatrix}$$

$$(4.2)$$

Concordia is formed by this d–q current. Harmonic assessment using Concordia is done in Park plane. In this plane harmonic-free stator current forms exact circular Concordia, but the presence of harmonics brings cleavages in this Concordia.

### 4.5.2 Pattern Generation and Inference

Concordia is generated for system containing different orders of harmonics. Figure 4.2 shows three patterns formed by d axis and q axis currents. The pattern, free from cleavage, corresponds to a balanced harmonic-free system. The pattern with four cleavages corresponds to system with fifth-order harmonic and the other pattern with five cleavages corresponds to system with sixth-order harmonic.

### 4.5.3 CMS Rule Set

For the analysis of such Concordia CMS rule set has been developed for identifying the presence of the single harmonic which may be present in the current signature [4, 5].

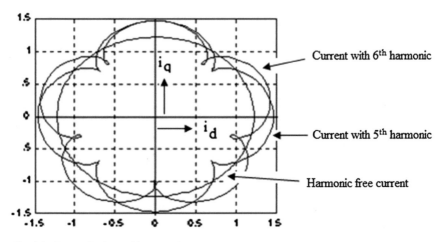

**Fig. 4.2** Concordia formed in Park plane

Rule 1  If cleavage appears, then there is harmonic in the system

Rule 2  If rule 1 is true and if the number of cleavages is C and order of harmonic is $n$, then $n = C + 1$

Rule 3  If above rule 1 is true, then there will be at least one cleavage at an angle $\alpha_n$ given by $\alpha_n = 270° + \theta/2$ for odd order ($n$) and $\alpha_n = 270° + \theta$ for even order ($n$), where $\theta = 360°/C$

Rule 4  If rules 1, 2, and 3 are true, then percentage of harmonic is proportional to depth of locus at the minima of the cleavages.

## 4.6  Experimentation

The block diagram of the experimental setup is shown in Fig. 4.3. For detection of fault current signals are to be captured through Hall probe sensors. Signals may also be taken through current transformer and rheostat arrangement signals are sampled, digitized, and collected in central processing unit with the help of data acquisition system.

*Example 4.1* Two induction motors of same rating, as shown in Table 4.1, are used for experimentation; first one is healthy motor and the second one is with broken rotor bars termed as faulty motor. Setup with faulty motor is shown in Fig. 4.4. These motors are started through direct on line (DOL) starters. 3-phase, 110 V, 50 Hz supply is provided to each of the motors. For capturing current signature, Hall probe is used. Transient current signals are captured at a sampling frequency of 5120 sample/sec., in a high-speed data acquisition system (DAS). Specifications of Hall probe and DAS are given in Table 4.2.

Details about fault detection are discussed in this section with the help of results and observation obtained by experimentation with this Example 4.1.

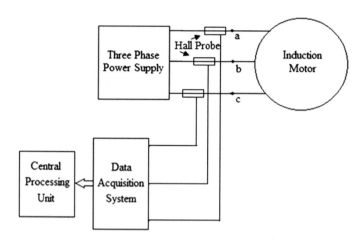

**Fig. 4.3**  Block diagram of the experiment

**Table 4.1**  Specification of motors

| Motor | Make | Cat No./Model | Power | Voltage | Frequency | Speed |
|-------|------|---------------|-------|---------|-----------|-------|
| Healthy | Marathon electric | D391/DVC 56T34F5301JP | 1/3 HP | 415 V | 50 Hz | 2980 rpm |
| Faulty | Marathon electric | D391/DVC 56T34F5301JP M-BRB | 1/3 HP | 415 V | 50 Hz | 2980 rpm |

**Fig. 4.4**  Experimental setup (*courtesy* Power System Lab, Department of Applied Physics, University of Calcutta, India)

**Table 4.2**  Specifications of Hall probe and data acquisition system

| Hall probe | | | Data acquisition system | | |
|------|-------|------------|------|-------|-------|
| Make | Model | Conversion | Make | Model | Speed |
| LEM | PR30 ACV 600 V CATIII | 30A ac/3 V ac | OROS | OR35, 8 channels | 100 mbps |

## 4.7 Concordia in Park Plane with Transient Signals at Starting

Starting currents are captured during the direct on line (DOL) starting of both healthy motor and motor with broken rotor bar (as mentioned in Example 4.1. Then Concordia of the current signals is formed for both the motors as shown in Table 4.3. It is observed that each Concordia consists of many cleavages. According to the rule set, these cleavages are for harmonics present in the stator current. It is also observed that all Concordia are different from each other. This suggests that harmonics present in the starting current of healthy motor and motor with broken bar rotor are different. However, from these complicated Concordia it is very difficult to gather sufficient information about harmonics.

**Table 4.3** Stator current Concordia in Park plane

| | |
|---|---|
| In Park plane stator current Concordia of healthy motor at starting | |
| In Park plane stator current Concordia of motor with broken rotor bar at starting | |

## 4.8 Radar of Starting Transients at Starting

Radar analysis is performed corresponding to these two sets of Concordia by developing radar diagrams as shown in Fig. 4.5. There are three approximate circles of which the innermost circle corresponds to the healthy motor and the intermediate circle corresponds to the motor with broken rotor bar. [Here the outermost circle corresponds to the motor with mass unbalanced rotor which will be discussed in consecutive chapter.] Thus radar analysis suggests that the increase of average diameter from normal diameter first indicates broken bar fault [and more increase of this diameter indicates rotor mass unbalance in the motor]. Same experimentation is done at different operating conditions with the two motors mentioned in Table 4.1. Some observations are presented in Table 4.4. It shows the same effect on average diameter, i.e., average diameter of the circle referred for broken rotor bar $[(D_{av})_{\text{broken-bar}}]$ is greater than that for healthy motor $[(D_{av})_{\text{healthy}}]$. [In next chapter of this book it will be shown that $(D_{av})_{\text{mass unbalance}} > (D_{av})_{\text{broken-bar}} > (D_{av})_{\text{healthy}}$.]

### 4.8.1 Identification of Broken Rotor Bar Fault

Stator current Concordia in Park plane presented in Table 4.3 gives significant change for broken rotor bar from healthy motors. Radar analysis from Fig. 4.5 suggests that the increase of average diameter first indicates broken rotor bar fault and more increase of this diameter indicates rotor mass unbalance in the motor.

## 4.9 Algorithm for Concordia and Radar-Based Diagnosis of Broken Rotor Bar

To find out the existence of faults like broken rotor bar in an induction motor, Concordia is formed in Park plane using motor starting current. Then radar analysis is done to classify the abnormality and an algorithm has been developed. Fault has been identified using radar analysis of stator current without focusing on a particular frequency or harmonics.

Based on the discussion, an algorithm has been developed for identification of broken rotor bar fault as follows:

(a) Stator current is to be captured in a data acquisition system.
(b) Concordia is to be formed, in Park plane.
(c) If the Concordia has any cleavage, then there are some harmonics in stator current.
(d) Concordia is to be compared with that of the motor under healthy condition.

**Table 4.4** Average diameter of radar of stator current at different loading conditions

| Condition | Radar diagram | Findings |
|---|---|---|
| Rated supply voltage at rated load | | $(D_{av})$broken-bar $> (D_{av})$healthy |
| Rated supply voltage at 50 % load | | $(D_{av})$broken-bar $> (D_{av})$healthy |

(continued)

**Table 4.4** (continued)

| Condition | Radar diagram | Findings |
|---|---|---|
| Rated supply voltage at no load | | $(D_{av})$broken-bar $> (D_{av})$healthy |

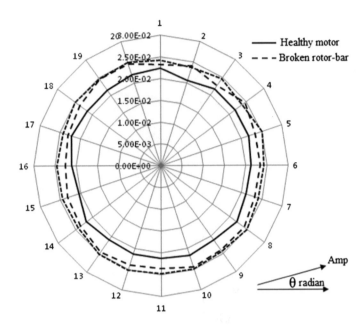

**Fig. 4.5** Radar diagrams corresponding to starting current of healthy motor and motor with broken rotor bar in polar coordinate (*note* diagnosis of rotor mass unbalance will be discussed in Chap. 5)

(e) If the comparison shows any difference, then radar analysis is to be performed. Measuring the diameter of the radars, conclusion can be drawn regarding the health of the motor, i.e., whether the motor is healthy or having fault with broken rotor bar.

## 4.10   Diagnosis Through Envelope Analysis of Motor Startup Current Using Hilbert and Wavelet Transform [6]

Radar-based assessment is capable of detecting broken rotor bar. However, main limitation of this approach is that radar-based assessment is unable to give detailed information related to the harmonics produced due to broken rotor bar. In this section attempt has been taken to overcome this limitation.

## 4.10.1 *Instantaneous Frequency and Hilbert Transform*

The concept of instantaneous frequency is introduced for transient/nonstationary signal analysis where the periodic frequency associated with some sinusoidal function loses its effectiveness. It is a time-varying parameter which defines the spectral peaks varying with time. It has significant practical importance in radar, communication, and biomedical application. It has only useful meaning for mono-component signal or narrowband of frequencies. Originally, it was defined in context of FM modulation theory. Instantaneous frequency is defined as $f_i = (1/2\pi)$ $d\Phi/dt$, i.e., phase variability. Again any AM/FM wave can be represented in complex form m.e $^{j\Phi t}$ as seen in [7]. To make an useful meaning of instantaneous frequency, an unique complex representation of a signal say $s(t)$ is obtained using Hilbert transform whether or not it corresponds to any physical reality and it can be treated like periodic sinusoidal frequency as described [7]. The complex signal is known as analytic signal. If the real-time signal is $s(t)$ and its Hilbert transform is $H[s(t)]$, then unique complex representation of the signal according to Gabor and Ville may be written as

$$z(t) = s(t) + j[H[s(t)]] \tag{4.3}$$

The Hilbert transform is defined as

$$H[s(t)] = \frac{1}{\pi} \int_{-\infty}^{\infty} \frac{x(\tau)}{(t-\tau)} d\tau \tag{4.4}$$

Using the mean value theorem, it can be evaluated as

$$H[s(t)] = \frac{1}{\pi t} \otimes s(t) \tag{4.5}$$

Therefore, $H[s(t)]$ is obtained from the convolution of the function $1/(\pi t)$ with the original function $s(t)$.

$H[s(t)]$ and $s(t)$ are supposed to be in quadrature because in theory they are out of phase by $\pi/2$, i.e., Hilbert transform is equivalent to the positive frequencies from the spectrum of $s(t)$ shifted by—$\pi/2$ and their amplitudes are doubled and the negative frequencies are removed. The Hilbert transform can be viewed as a filter which has the property to eliminate the negative frequencies and retain the positive frequencies with their phase shift of $\pi/2$. The complex signal, $z(t)$, is known as the analytic signal which does not always correspond to the signal and its quadrature. When there may be significant leakages from positive spectral components into the negative spectral region, the HT will not produce quadrature component of input signal. Under such condition, analysis through this HT may lead to confusing results. If the signal is of the form $\alpha(t)\cos\Phi(t)$ which is like a real FM signal, then it may be written as

$$z(t) = s(t) + jH[s(t)] = \alpha(t) \cos \Phi(t) + jH[\alpha(t) \cos \Phi(t)]$$
$$\text{or, } z(t) = \alpha(t)[\cos \Phi(t) + j \sin \Phi(t)] \tag{4.6}$$
$$\text{or, } \quad z(t) = \alpha(t)e^{j\Phi(t)}$$

Now the analytic signal is of frequency modulated form as given in (3.6). However, if the spectra of $\alpha(t)$ and $\Phi(t)$ are not separately considered, the HT will produce overlapping and phase-distorted functions.

For meaningful practical application, the amplitude spectra of $\alpha(t)$ and the phase spectra of $\Phi(t)$ are considered separately and the amplitude spectra of $\alpha(t)$ correspond to low-frequency zone of the system, whereas the phase spectra of $\Phi(t)$ occupy high-frequency portion. In this paper, the discussion is limited to low-frequency amplitude spectra.

The analytic signal will be accurate complex representation of the real signal $s(t)$ for narrowband amplitude spectra of $\alpha(t)$ only when the real signal, $s(t)$, and its HT are in quadrature resulting in better estimation of instantaneous frequency.

### 4.10.2  Envelope Detection

The envelope of the signal, $s(t)$, is defined as modulus of analytic signal $z(t)$ given by

$$E(t) = |s(t) + jH[s(t)]|$$
$$\text{or } E(t) = \alpha(t) \tag{4.7}$$

The analytic signal contains an amplitude component and phase component generally, of which our interest is focused in low-frequency zone of the analytic signal $z(t)$, i.e., in the envelope $\alpha(t)$. It is a new dimension for detection of induction motor fault from the spectrum analysis of the envelope signal. The spectrum analysis of the envelope yields better detection ability of fault than the spectrum analysis of the original signal as the power frequency is almost removed from the signal.

### 4.10.3  Methodology

A method has been introduced for the extraction of time-varying low-frequency oscillation from the induction motor startup current. The concept of instantaneous frequency has been introduced here to search for the spectral peaks which vary with time. It has only meaningful application for signals having narrow range of frequencies. The Hilbert transform is used here to generate complex analytic signal from an input signal, the motor starting current which has single harmonic (50 Hz) surroundings. If $s(t)$ is the signal and $H[s(t)]$ is its Hilbert transform, the analytic signal $z(t)$, given in (4.3), accurately represents the original signal in its complex form only when $s(t)$ and $H[s(t)]$ are in quadrature. However, the components of $z(t)$ are not always orthogonal because of leakage from positive frequencies to negative frequencies of the signal. The analysis using instantaneous frequency like the periodic frequency is possible when the analytic signal will be of the form of either FM or AM signal. As startup current is sinusoidal and may be written as $\alpha(t)\cos\Phi(t)$, then the analytical signal will be expressed in the form of $z(t) = \alpha(t)e^{j\Phi(t)}$ as given in (4.6) which is like FM signal. The analytical signal has two parts amplitude and phase. $\alpha(t)$ is the modulus of the analytic signal given by $E(t) = |\,s(t) + jH[s(t)]\,|$, where $E(t)$ being the envelope of the signal. The effect of supply frequency has been removed and this makes the analysis most effective. The amplitude spectra belong to the low-frequency portion of the motor current signal which is used in this paper for detection of broken rotor bar. For the meaningful application of instantaneous frequency, the processed signal, i.e., the amplitude part of the analytic signal, $\alpha(t)$, needs to be filtered in the narrowband zone. Higher order wavelet is the excellent choice to filter the narrowband frequencies below power supply frequency. In its higher wavelet level frequency bands are shown in Table 4.5.

In the present work, higher level detailed coefficients, using mother wavelet db10, at the levels 8th, 9th, and 10th, the frequency bands are very narrow and below 50 Hz. At higher level of decomposition, overlapping of adjacent bands disappears, minimizing the possible leakage. The method of using Hilbert transform along with DWT presents an excellent method for detection of broken rotor bar fault using transient startup current, identifying the low-frequency oscillation.

**Table 4.5** Spectral frequency band at different decomposition levels

| Decomposition details | Frequency bands (Hz) |
|---|---|
| Detail level 1 | 2048–4096 |
| Detail level 2 | 1024–2048 |
| Detail level 3 | 512–1024 |
| Detail level 4 | 256–512 |
| Detail level 5 | 128–256 |
| Detail level 6 | 64–128 |
| Detail level 7 | 32–64 |
| Detail level 8 | 16–32 |
| Detail level 9 | 8–16 |
| Detail level 10 | 4–8 |

## *4.10.4  Discrete Wavelet Transform-Based Assessment*

The motor current signatures for both no load and partial load conditions are captured. Then signal envelopes (from the data of Example 4.1) are obtained which are shown in Figs. 4.6, 4.7, 4.8, and 4.9. The signal envelopes are the absolute value of the analytical signal obtained on performing the Hilbert transform (HT) of the original signature. The signal envelopes are decomposed into detailed and approximate coefficients through DWT using "db10" of Daubechies family up to wavelet level 10. Each detail is then reconstructed.

Absolute reconstructed details of the starting current signals and their envelopes both for the faulty and healthy motor at no load and at partial load are shown in Figs. 4.10, 4.11, 4.12, 4.13, 4.14, 4.15, 4.16, and 4.17.

The whole analysis is performed using Matlab. The objective is to extract lower side harmonics below 50 Hz for which higher order wavelet "db10" at higher levels 8th, 9th, and 10th are utilized. The statistical parameters, namely mean and standard deviation of the absolute reconstructed detailed coefficients of the starting current signals and their respective signal envelopes for both the motors at no load and partial load, are estimated. These are considered as fault parameters. From Tables 4.6 and 4.7, it is observed that these values at 7th level for the original signal

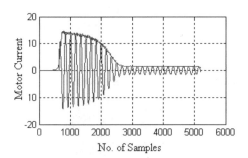

**Fig. 4.6**  Current signal and its envelope for healthy motor at no load

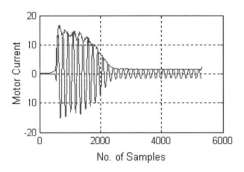

**Fig. 4.7**  Current signal and its envelope for faulty motor at no load

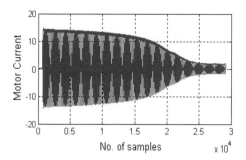

**Fig. 4.8** Current signal and its envelope for healthy motor at partial load

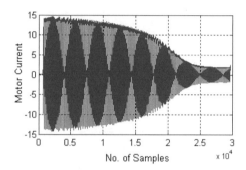

**Fig. 4.9** Current signal and its envelope for faulty motor at partial load

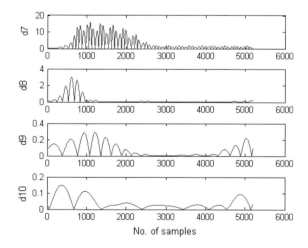

**Fig. 4.10** Reconstructed details of current signal at no load for healthy motor

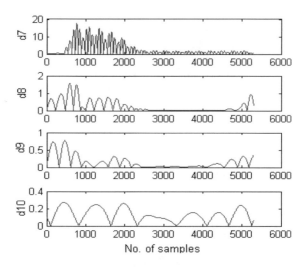

**Fig. 4.11** Reconstructed details of current signal at no load for faulty motor

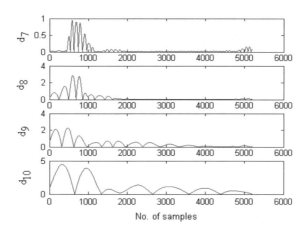

**Fig. 4.12** Reconstructed details of current signal envelope at no load for healthy motor

are much higher than the corresponding values for the signal envelope which ensures that the power frequency is not present in the signal envelopes. This makes the detection easier and cleaner as the lower side harmonics are free from the spectral leakage effect of 50 Hz. The DWT of the original signals are based on Fourier sinusoidal frequency, whereas DWT of the signal envelopes works on the concept of instantaneous oscillating frequency. In comparison to the original signal, the envelope analysis provides better detection ability as it is observed from the curves in Figs. 4.18, 4.19, 4.20, 4.21 and Tables 4.6 and 4.7.

The curves representing the statistical parameters—mean and standard deviation —of the absolute reconstructed detailed coefficients of the original signal lie below

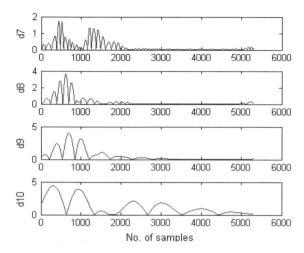

**Fig. 4.13** Reconstructed details of current signal envelope at no load for faulty motor

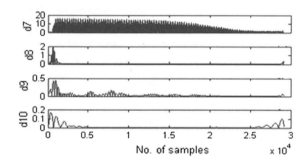

**Fig. 4.14** Reconstructed details of current signal at partial load for healthy motor

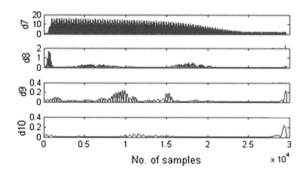

**Fig. 4.15** Reconstructed details of current signal at partial load for faulty motor

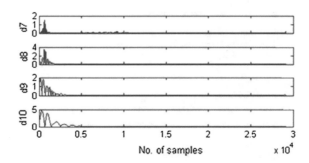

**Fig. 4.16** Reconstructed details of current signal envelope for healthy motor under partial load condition

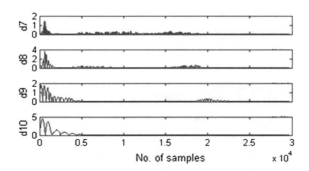

**Fig. 4.17** Reconstructed details of current signal envelope for faulty motor under partial load condition

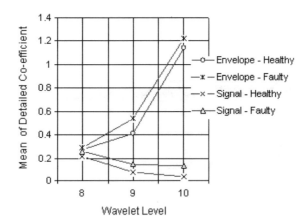

**Fig. 4.18** Mean of reconstructed details of stator current at no load

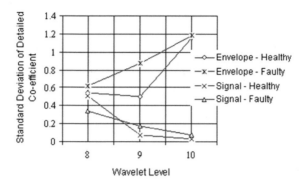

**Fig. 4.19** Standard deviation of reconstructed details of stator current at no load

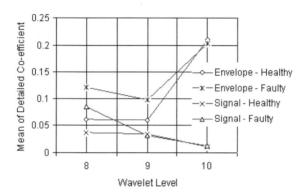

**Fig. 4.20** Mean of reconstructed details of stator current at partial load

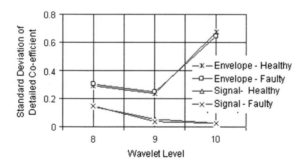

**Fig. 4.21** Standard deviation of reconstructed details of stator current at partial load

the curves representing those parameters corresponding to signal envelopes except at 8th wavelet level; the only mean corresponding to the original signal for the faulty motor under partial load being higher value than the mean corresponding to

**Table 4.6**  Statistical parameters of absolute reconstructed details of motor current signal

| Load condition | Motor condition | Mean of absolute signal at wavelet level | | | | Standard deviation of absolute signal at wavelet level | | | |
|---|---|---|---|---|---|---|---|---|---|
| | | 7 | 8 | 9 | 10 | 7 | 8 | 9 | 10 |
| No load | Healthy | 2.973 | 0.2106 | 0.07209 | 0.03979 | 3.686 | 0.5152 | 0.07774 | 0.03629 |
| | Faulty | 2.821 | 0.2577 | 0.1469 | 0.1272 | 3.632 | 0.3432 | 0.1746 | 0.07431 |
| Partial load | Healthy | 5.485 | 0.03797 | 0.03507 | 0.01162 | 4.302 | 0.1446 | 0.05344 | 0.02548 |
| | Faulty | 5.3 | 0.0846 | 0.03208 | 0.01276 | 4.249 | 0.1484 | 0.04345 | 0.02636 |

**Table 4.7**  Statistical parameters of absolute reconstructed details of motor current signal envelope

| Load condition | Motor condition | Mean of absolute signal envelope at wavelet level | | | | Standard deviation of absolute signal envelope at wavelet level | | | |
|---|---|---|---|---|---|---|---|---|---|
| | | 7 | 8 | 9 | 10 | 7 | 8 | 9 | 10 |
| No load | Healthy | 0.07287 | 0.271 | 0.4057 | 1.14 | 0.1616 | 0.5472 | 0.5051 | 1.158 |
| | Faulty | 0.1982 | 0.2883 | 0.5345 | 1.226 | 0.3199 | 0.6174 | 0.8693 | 1.186 |
| Partial load | Healthy | 0.03936 | 0.06242 | 0.0603 | 0.2097 | 0.08933 | 0.2903 | 0.2361 | 0.6743 |
| | Faulty | 0.06831 | 0.1219 | 0.09776 | 0.2019 | 0.1075 | 0.3018 | 0.2495 | 0.6459 |

the signal envelope of the healthy motor but still its value is less than the mean corresponding to the signal envelope of the faulty motor. This confirms that the fault parameters corresponding to instantaneous oscillating frequency of the signal envelopes have higher magnitude than those of the original signal, resulting in higher detection ability and better representation of the nonstationary starting current of the motors both at no load and partial load. The signal envelope at no load produces higher values of the statistical mean and standard deviation of the absolute reconstructed details for faulty motor at 8th, 9th, and 10th wavelet levels than those of the healthy motor. The 9th level indicates much larger change of the respective parameter for the faulty motor than those for the healthy motor at no load. The signal envelope at partial load gives higher values of statistical parameters —mean and standard deviation—of absolute reconstructed details for faulty motor at 8th and 9th levels, but less value at the 10th level than those for the healthy motor. Therefore, it is observed that the 9th level is most sensitive to detect rotor broken bar fault, both at partial load and no load. Thus the lower side harmonics below 50 Hz extracted from the signal envelopes is the most effective for detection of the rotor broken bar fault.

It can be inferred from the observations discussed above that the envelope analysis based on instantaneous frequency gives better description of transient starting current of motor through Hilbert transform and wavelet transform for detection of the rotor broken bar fault in an induction motor.

## 4.10.5   Advantages and Disadvantages

The main advantage of this method is that envelope detection using Hilbert transform-based instantaneous frequency for extraction of lower side harmonics below supply frequency makes diagnosis easier as because the power frequency is eliminated and the spectral leakage is minimized which improves detection ability. This method works with higher resolution due to choice of higher order wavelet at higher wavelet level to extract left-side harmonics. The main disadvantage is once the sampling frequency is selected, the bands become fixed which means some ranges of frequencies are unexplored.

## 4.11   Conclusion

In this chapter rotor broken bar fault is first assessed by radar analysis with phase current, wherefrom it is very difficult to detect if there is any broken bar or not. Then radars are formed in Park plane wherefrom the fault can be differentiated easily compared with those of normal motor. Then harmonics generated due to rotor broken bar has been assessed using Hilbert and wavelet transforms. The main contribution in this is the application of new concept, envelope analysis, to extract low-frequency oscillation. This is done by the use of Hilbert and wavelet transform. This method of diagnosis works with higher resolution. Apart from this, it can also handle short data effectively. The number of computation is also less and simple compared to FFT and other time–frequency techniques—this makes the method suitable for online industrial application. The method may be extended to diagnose other faults of stator and rotor asymmetry and also to research in the areas of low-frequency oscillations.

## References

1. Deleroi W (1984) Broken bars in squirrel cage rotor of an induction motor-part I: description by superimposed fault currents. Arch Elektrotech 67:91–99
2. Filippetti F, Franceschini G, Tassoni C, Vas P (1998) AI techniques in induction machines diagnosis including the speed ripple effect. IEEE Trans Ind Appl 34:98–108
3. Chattopadhyay S, Karmakar S, Mitra M, Sengupta S (2012) Radar analysis of stator current Concordia for diagnosis of unbalance in mass and cracks in rotor bar of an squirrel cage induction motor. Int J Model Measur Control Gen Phys Electr Appl AMSE Ser A 85(1):50–61. ISSN: 1259-5985
4. Chattopadhyay S, Mitra M, Sengupta S (2011) Electric power quality, 1st edn. Springer, New York
5. Chattopadhyay S, Mitra M, Sengupta S (2007) Harmonic analysis in a three-phase system using park transformation technique. AMSE Int J Model Simul Model-A 80(3):42–58

6. Ahamed SK, Karmakar S, Mitra M, Sengupta S (2011) Diagnosis of broken rotor bar fault of induction motor through envelope analysis of motor startup current using hilbert and wavelet transform. J Innovative Syst Des Eng 2(4):163–176
7. Boashash B (1992) Estimating and interpreting the instantaneous frequency of a signal—part 1: fundamentals. Proc IEEE 80(4):520–538

# Chapter 5
# Rotor Mass Unbalance

**Abstract** The chapter deals with rotor mass unbalance fault. Diagnosis is done both using steady-state signal as well as starting current transients. Concordia is formed using steady-state current and analyzed. The Concordia-based assessment is done using starting current transients. Then radars are formed in Park plane wherefrom the faults are differentiated compared with those of normal motor. Then harmonics generated due to rotor broken bar has been assessed using FFT and wavelet transforms.

**Keywords** Rotor mass unbalance · Concordia · Discrete wavelet transform · FFT · Park plane · Radar

**Chapter Outcome**

After completion of the chapter, readers will be able to gather knowledge and information regarding the following areas:

- Rotor mass unbalance
- Steady-state approaches for rotor mass unbalance assessment
- Diagnosis using starting transients
- Concordia-based assessment
- Radar-based diagnosis
- Diagnosis using wavelet transform.

## 5.1 Introduction

In this chapter, different methods for assessment of rotor mass unbalance fault have been discussed and results of tests performed in laboratory are depicted. First, a description of the fault, its causes, and effects and then the diagnosis of the fault using different signal processing tools are mentioned. Motor current signature analysis techniques as well as vibration analysis technique have been used for the

© Springer Science+Business Media Singapore 2016
S. Karmakar et al., *Induction Motor Fault Diagnosis*,
Power Systems, DOI 10.1007/978-981-10-0624-1_5

purpose of diagnosis. Finally, a comparative study among different techniques, theiroutcomes, advantages, and disadvantages are discussed in the conclusion of this chapter.

## 5.2   Rotor Mass Unbalance

Rotor mass unbalance is an uncommon fault in induction motor. It is one kind of mechanical-related fault. Different reasons for occurrence of this fault are as follows:

- manufacturing defect
- nonsymmetrical addition or subtraction of mass around the center of rotation of rotor
- internal misalignment of rotor and stator
- motor shaft bending.

For any of the above reasons the weight distribution axis of the rotor does not coincide with the rotational axis of the rotor and hence axis of rotation of the rotor does not coincide with the geometrical axis of the stator resulting in air-gap eccentricity in the motor. For this reason rotor mass unbalance fault is considered as a rotor fault. Points to remember here that in a healthy motor, rotor is centrally aligned with the stator and the axis of rotation of the rotor is the same as the geometrical axis of the stator. This results in identical air gap between the outer surface of the rotor and the inner surface of the stator. However, if the air-gap eccentricity occurs, air gap between the outer surface of the rotor and the inner surface of the stator will not be uniform. Due to this fault in an induction motor, electromagnetic pull on the rotor will be unbalanced and in severe case of rotor eccentricity, due to unbalanced electromagnetic pull, rotor may even rub the inner surface of the stator. Rotor mass unbalance faults are of three types: (i) static unbalance, (ii) dynamic unbalance, and (iii) couple unbalance.

When an induction motor with rotor mass unbalance fault rotates, air-gap length oscillates which causes variation in air-gap flux density with some harmonics and hence variation in induced voltage in the winding results. These cause variation in stator current frequency along with some harmonics given by (5.1).

$$f_{\mathrm{ubm}} = f \left[ \frac{k(1-s)}{p} + 1 \right] \tag{5.1}$$

where $f$ is the supply frequency, $s$ is the slip of the motor, $p$ is the number of pole pair, and $k$ is an integer = 1, 2, 3, 4,...

Different techniques of diagnosing mass unbalance fault in an induction motor using different signal processing tools are described in the following sections.

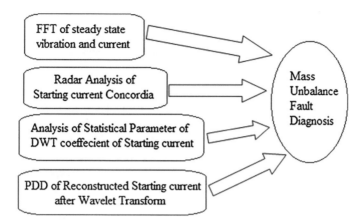

**Fig. 5.1**  Diagnosis of rotor mass unbalance fault using different techniques

## 5.3  Mathematical Tools Diagnosis of Rotor Mass Unbalance

For diagnosis of rotor mass unbalance, both transient and steady-state stator current are useful. Different techniques used for rotor mass unbalance are shown in Fig. 5.1. First, the conventional method FFT is performed on the steady-state stator current and vibration signals of the induction motor. Then analysis is performed on the transient part of the motor current, i.e., starting current. Radar analysis and discrete wavelet transform (DWT) are performed on this starting current. After performing DWT, different statistical parameters and power detail density (PDD) of the DWT coefficients are determined to diagnose mass unbalance fault. The methods have been described in subsequent subsections.

## 5.4  FFT-Based Diagnosis of Steady-State Motor Vibration and Current Signatures [1]

Individual faults are sensitive to particular harmonics and the amplitude of the harmonics changes for different faults. Also, if the amplitude of these harmonics is smaller by more than 50 dB, compared to the fundamental frequency components' amplitude, then the faulty motor cannot be differentiated from the healthy one [2].

In case of induction motor with unbalance rotor, air-gap flux changes during the run of the motor. As a result induced voltage in the rotor changes which changes the stator voltage and hence change in stator current. This change in stator current depends on the amount of variation of the air-gap flux. It is proportional to the flux density squared waveform in the induction motor [3]. Thus it can be concluded that motor current should give a reflection of the mass unbalance fault in the motor.

Effect of mass unbalance fault is also associated with change in vibration for which generally, by the analysis of vibration spectrum, motor fault is detected. In this work, both the current signature and vibration signature are captured and analyzed to diagnose the rotor mass unbalance fault.

*Example 5.1* Fault diagnosis will be discussed using this example in this chapter. Experiment is carried out on test setup manufactured by Spectra Quest, USA, having a high-speed data acquisition system. The experimental setup is shown in Fig. 5.2. Ratings of a healthy induction motor and another induction motor with mass unbalance fault, which are used for the experiment, are shown in Table 5.1. Motor is loaded mechanically and is supplied through power supply converter as shown in the block diagram in Fig. 5.3. For current signature Hall probe and for vibration signature, piezoelectric transducers are used. Specifications of the transducers are shown in Table 5.2. Current data of all the three phases and for vibration data in *X, Y,* and *Z* directions are collected.

Details about fault detection are discussed in the subsequent sections of the chapter with the help of results and observation obtained by experimentation with this Example 5.1.

**Fig. 5.2** Experimental setup (*courtesy* Power System Lab, Department of Applied Physics, University of Calcutta, India)

**Table 5.1** Specifications of motors

| Motor | Make | Cat no./model | Power (HP) | Voltage (V) | Frequency (Hz) | Speed (rpm) |
|---|---|---|---|---|---|---|
| Healthy | Marathon electric | D391/DVC 56T34F5301JP | 1/3 | 415 | 50 | 2980 |
| Faulty | Marathon electric | D391/DVC 56T34F5301JP M-UBM | 1/3 | 415 | 50 | 2980 |

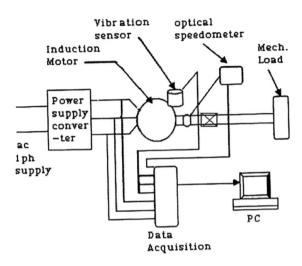

**Fig. 5.3** Block diagram for the experiment

**Table 5.2** Specifications of the transducers

| Hall probe transducer | | | Piezoelectric transducer | | |
|---|---|---|---|---|---|
| Make | Specifications | Conversion | Make | Specifications | Conversion |
| LEM | PR30 ACV 600 V CATIII | 30 A ac/3 V ac | IMI | 604B31 ICP | 9.6 V/(mm/s$^2$) |

## 5.5 Current Signature Analysis

Current data for all the three phases are collected but as all are similar, only those for $Y$ phase are presented here. In Table 5.3 the amplitudes of the predominant harmonics, obtained on performing the FFT of the signal, are given for $Y$ phase for both healthy and faulty motors of same rating. Graphs of the FFT of the current signals are shown in Figs. 5.4, 5.5, 5.6, 5.7, 5.8, and 5.9.

**Table 5.3** Amplitude of current harmonics of healthy and faulty motor

| Spectral freq. (Hz) | Order of harmonics | Amplitude in mA | | | | | |
|---|---|---|---|---|---|---|---|
| | | Healthy motor | | | Rotor unbalanced motor | | |
| | | No load | Single mass load | Double mass load | No load | Single mass load | Double mass load |
| 50 | 1st | 124.6 | 125 | 126.6 | 118.6 | 118.8 | 121 |
| 100 | 2nd | 2.615 | 2.544 | 2.7 | 3.442 | 2.7 | 2.7 |
| 150 | 3rd | 3.636 | 3.701 | 3.782 | 2.935 | 3.03 | 2.8 |
| 200 | 4th | 1.12 | 1.125 | 1.153 | 1.232 | 1.174 | 1.2 |
| 250 | 5th | 1.778 | 1.809 | 1.83 | 1.754 | 1.566 | 1.743 |
| 350 | 7th | 1.356 | 1.494 | 1.586 | 1.407 | 1.41 | 1.65 |

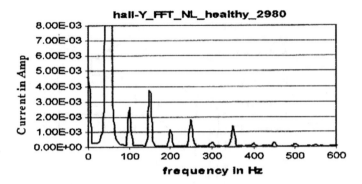

Fig. 5.4   FFT of current of $Y$ phase of healthy motor under no load

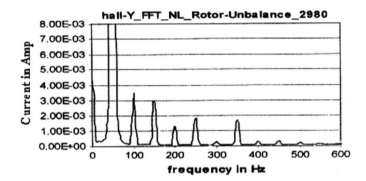

Fig. 5.5   FFT of current of $Y$ phase of rotor mass unbalance motor under no load

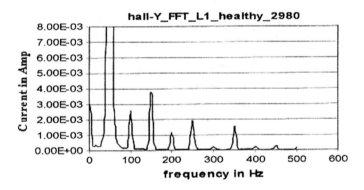

Fig. 5.6   FFT of current of $Y$ phase of healthy motor with partial (single mass) load

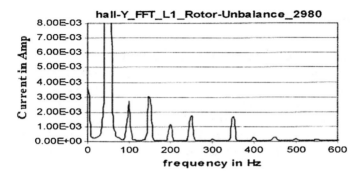

**Fig. 5.7** FFT of current of Y phase of rotor mass unbalance motor with partial (single mass) load

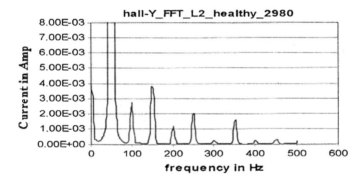

**Fig. 5.8** FFT of current of Y phase of healthy motor with double mass load

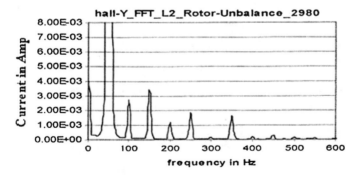

**Fig. 5.9** FFT of current of Y phase of rotor mass unbalance motor with double mass load

## 5.6   Vibration Analysis

Data for all the three directions $X$, $Y$, and $Z$ are collected but vibration in the $Z$ direction being more prominent these are considered for analysis. FFT of this vibration is taken for both healthy and faulty motors under no load and load condition, and these are shown in Figs. 5.10, 5.11, 5.12, 5.13, 5.14, and 5.15.

In Table 5.4, amplitudes of FFT of different vibration harmonics for both healthy and faulty motors under load and no load conditions are shown. Distinct variations compared to the healthy motor are observed in case of the motor with mass unbalance rotor.

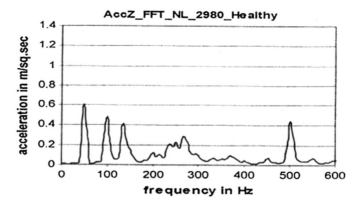

**Fig. 5.10** FFT of vibration signal of healthy motor at no load

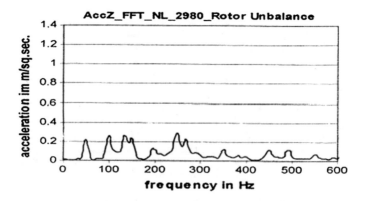

**Fig. 5.11** FFT of vibration signal of rotor mass unbalance motor at no load

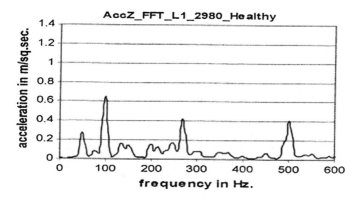

**Fig. 5.12** FFT of vibration signal of healthy motor with single mass load

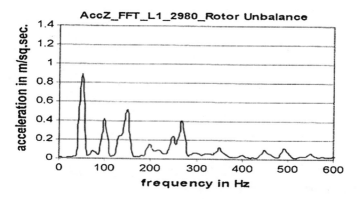

**Fig. 5.13** FFT of vibration signal of rotor mass unbalance motor with single mass load

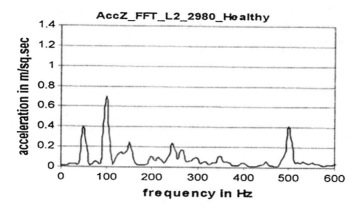

**Fig. 5.14** FFT of vibration signal of healthy motor with double mass load

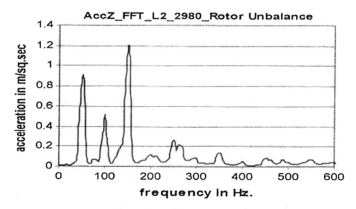

**Fig. 5.15** FFT of vibration signal of rotor mass unbalance motor with double mass load

**Table 5.4** Amplitude of vibration harmonics of healthy and faulty motor

| Harmonic order | Amplitude in m/s$^2$ | | | | | |
|---|---|---|---|---|---|---|
| | Healthy motor | | | Mass unbalance rotor motor | | |
| | At no load | Single mass load | Double mass load | At no load | Single mass load | Double mass load |
| 1st | 0.604 | 0.267 | 0.392 | 0.218 | 0.888 | 0.914 |
| 2nd | 0.481 | 0.644 | 0.692 | 0.259 | 0.421 | 0.517 |
| 3rd | 0.159 | 0.141 | 0.246 | 0.229 | 0.509 | 1.198 |

## 5.7   Concordia-Based Diagnosis by Steady-State Stator Current in Park Plane [4]

Steady-state current is a useful signal for fault diagnosis as it is available for long period of time which makes data capturing easier than transient period during switching on or off. In this section, Concordia-based fault diagnosis has been done using steady-state motor current in Park plane.

### 5.7.1   Concordia in Park Plane

Stator currents are transformed into direct–quadrature (*d–q*) plane by Park matrix as follows:

$$\begin{pmatrix} i_{\mathrm{d}} \\ i_{\mathrm{q}} \end{pmatrix} = \sqrt{\frac{2}{3}} \times \begin{pmatrix} \cos\theta & \cos\left(\theta - 2\pi/3\right) & \cos\left(\theta - 4\pi/3\right) \\ -\sin\theta & -\sin\left(\theta - 2\pi/3\right) & -\sin\left(\theta - 4\pi/3\right) \end{pmatrix} \times \begin{pmatrix} i_a \\ i_b \\ i_c \end{pmatrix}$$

(5.2)

Concordia formed by these currents under healthy condition of the motor looks circular in Park plane. In this plane, harmonic-free stator current form exact circular Concordia, but the presence of harmonics brings cleavages in this Concordia. Harmonics assessment using Concordia is done in [5, 6] in Park plane.

### 5.7.2 Pattern Generation and Inference

Concordia is generated for system containing different orders of harmonics. Figure 5.16 shows three patterns formed by $d$ axis and $q$ axis currents. The pattern, free from cleavage, corresponds to a balanced harmonic-free system. The pattern with four cleavages corresponds to system with fifth-order harmonic and the other pattern with five cleavages corresponds to system with sixth-order harmonic.

### 5.7.3 CMS Rule Set

For analysis of such Concordia CMS rule set has been used for identifying the presence of the single harmonic present in the system [5, 6].

**Rule 1:** If cleavage appears, then there is harmonic in the system
**Rule 2:** If rule 1 is true and if the number of cleavages is $C$ and order of harmonic is $n$, then $n = C + 1$
**Rule 3:** If rule 1 is true, then there will be at least one cleavage at an angle given by $\alpha_n = 270° + \theta/2$ for odd order ($n$) and $\alpha_n = 270° + \theta$ for even order ($n$), where $\theta = 360°/C$
**Rule 4:** If rules 1, 2, and 3 are true, then percentage of harmonic is almost proportional to depth of locus at the minima of the cleavages.

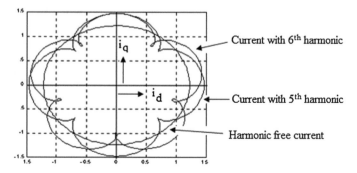

**Fig. 5.16** Concordia formed in Park plane

**Table 5.5**  Stator current Concordia in Park plane

| Stator current Concordia of healthy motor at starting in Park plane | 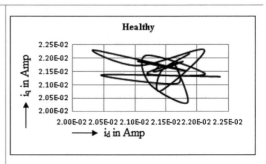 |
|---|---|
| Stator current Concordia at starting of motor with mass unbalance fault in Park plane | 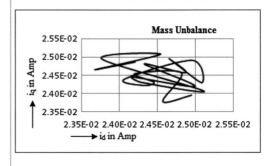 |

## 5.8  Concordia-Based Assessment of Stator Current in Park Plane at Starting

During direct on line (DOL) starting of motors, stator currents are captured for both healthy motor and motor with rotor mass unbalance fault. Then Concordia of these currents are formed as shown in Table 5.5. It is observed that each Concordia consists of many cleavages. According to the rule set, these cleavages are for harmonics present in the stator current. It is also observed that all Concordia are different from each other. This suggests that harmonics present in the starting current of healthy motor and motor with mass unbalance fault are different.

## 5.9  Radar Analysis of Stator Current in Park Plane at Starting

Concordia patterns described in the previous section show difference at fault condition from normal. However, analytically as the Concordia patterns are complex in such case, it has become very difficult to assess those difference. This limitation has been overcome by radar analysis described in the following subsection.

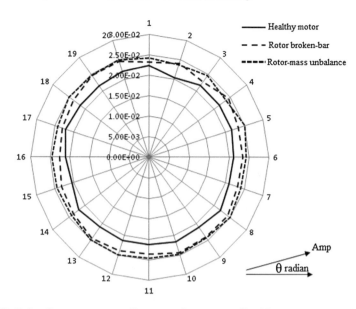

**Fig. 5.17** Radar diagrams corresponding to starting current of healthy motor, motor with broken rotor bar and motor with mass unbalance

### 5.9.1   Radars in Park Plane

Corresponding to the above two sets of Concordia, radar analysis is performed by developing radar diagrams as shown in Fig. 5.17. In this figure there are three approximate circles of which the innermost circle corresponds to the healthy motor and the outermost circle corresponds to the motor with rotor mass unbalance fault. (Here the intermediate circle corresponds to the motor with broken rotor bar which has been already discussed in the fourth chapter of this thesis.) Thus radar analysis suggests that the increase of average diameter from normal diameter first indicates broken bar fault and more increase of this diameter indicates rotor mass unbalance in the motor. Same experimentation is done at different operating conditions with the two motors mentioned in Table 5.1. Some observations are presented in Table 5.6. It also shows the same effect on average diameter, i.e., $(D_{av})_{\text{mass unbalance}} > (D_{av})_{\text{broken-bar}} > (D_{av})_{\text{healthy}}$.

### 5.9.2   Algorithm for Assessment of Rotor Mass Unbalance and Broken Rotor Bar Faults

Stator current Concordia in Park plane, presented in Table 5.6, shows significant variations in Concordia for mass unbalance and broken bar from healthy motor. Radar

**Table 5.6** Average diameter of radar of stator current at different conditions

| Condition | Radar | Findings |
|---|---|---|
| Rated supply voltage at rated load | | $(D_{av})_{\text{mass unbalance}} > (D_{av})_{\text{broken-bar}} > (D_{av})_{\text{healthy}}$ |
| Rated supply voltage at 50 % load | | $(D_{av})_{\text{mass unbalance}} > (D_{av})_{\text{broken-bar}} > (D_{av})_{\text{healthy}}$ |

(continued)

**Table 5.6** (continued)

| Condition | Radar | Findings |
|---|---|---|
| Rated supply voltage at no load | | $(D_{av})_{\text{mass unbalance}} > (D_{av})_{\text{broken-bar}} > (D_{av})_{\text{healthy}}$ |

**Table 5.7** Comparison of statistical parameters of healthy and mass unbalance rotor motor

| Wavelet level | Motor condition | Statistical data | | | |
|---|---|---|---|---|---|
| | | Max–Min | Mean | STD | RMS |
| 4 | Healthy | 5.492 | −0.0168 | 0.756 | 0.7538 |
| | Faulty | 5.751 | 0.0135 | 0.801 | 0.7992 |
| 5 | Healthy | 11.84 | 0.0326 | 2.015 | 2.005 |
| | Faulty | 15.42 | 0.0647 | 2.39 | 2.3784 |
| 6 | Healthy | 153 | 0.6817 | 36.63 | 36.322 |
| | Faulty | 150.2 | −0.0821 | 38.34 | 38.012 |

diagrams shown in Fig. 5.17 suggest that the increase of average diameter first indicates rotor broken bar failure and more increase of this diameter indicates rotor mass unbalance in the motor. This is also supported, as presented in Table 5.7, when motors are run under different load conditions. Based on the above results, an algorithm has been developed for identification of mass unbalance and broken bar faults as follows:

a. Stepped down stator currents are to be captured by data acquisition system.
b. Concordia in Park plane are to be formed.
c. If Concordia cleavages are present, then there are some harmonics in stator current.
d. Concordia corresponding to the faulty motors is to be compared with that of the healthy motor.
e. If the comparison shows a different result, then radar analysis is to be performed. Comparing the average diameter of radar diagrams, it is concluded whether the motor is healthy or having the fault. Also, faults like rotor mass unbalance and broken rotor bar can be isolated from the radars.

## 5.10   Discrete Wavelet Transform-Based Fault Diagnosis Using Starting Current at No Load [7]

Nature of steady-state motor current depends on load condition, but starting current is less dependent on load condition. Nature of starting current largely depends on motor parameters. Thus starting current is very useful signal for motor fault diagnosis. In the following section, DWT-based fault diagnosis has been done using starting current at no load.

A. Mass unbalance

Rotor of a motor possesses some amount of residual unbalance, whatsoever well they are balanced, which develops eccentricity and centrifugal force with the rotation of the motor. In Fig. 5.18, the geometric center of rotor is at $P$ and its center of gravity $G$ is at a distance $\varepsilon$ called eccentricity. In the fixed $OXY$ axis system, the geometry of unbalance whirl at the rotor is shown.

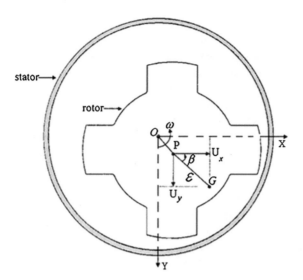

**Fig. 5.18** Unbalanced centrifugal force due to mass unbalance rotor

The mass unbalance $U$ is defined as the product of the unbalance mass $m$ and eccentricity $\varepsilon$. The centrifugal force $F_m$ in the $Y$–$Z$ plane, when the eccentric rotor is rotating at an angular speed of $\omega t$, can be expressed by the following equations [8]:

$$U_X = m\varepsilon \cos \beta \tag{5.3}$$

$$U_Y = m\varepsilon \sin \beta \tag{5.4}$$

$$F_m = \omega^2 U_Y \cos \omega t + \omega^2 U_X \sin \omega t \tag{5.5}$$

where $\beta$ is the angle of eccentricity with the horizontal.

Total unbalance force as function of time is the sum of the centrifugal force and unbalance electromagnetic force. Rotor eccentricity increases due to this unbalance force which tries to pull it further away from the stator bore center and this leads to variation of air-gap length. Unbalance force is changed by the rotational speed. In addition, vibration arising out of eccentricity which develops excessive stress on the bearing of the rotor deteriorates the performance of the motor and reduces the life.

During start, the unbalance force is very high where the rotor is pulled across the whole air gap. Further, this may lead to bent rotor shaft, bearing wear and tear. If not detected, it may cause stator to lead to rotor rub causing a major breakdown of the machine during start up.

Due to rotor unbalance, dynamic eccentricity results in oscillation in the air-gap length, causing air-gap flux density variation and voltage induced in the winding and hence currents are changed whose frequencies are determined by the frequency of the air-gap flux density harmonics.

The stator current harmonics is given by (5.1). Steady-state current signal is analyzed using FFT technique to search for the characteristic frequency due to fault which largely depends on load [3, 9]. At light load or no load for small- and medium-sized motor, fault harmonics are not easily detectable due to their closeness to fundamental and smaller amplitudes. But during the starting (when started through DOL starter), the stator current is very high, typically 8–10 times of steady-state current which contains much more distinct information regarding fault characteristics. Also, starting current is less sensitive to motor loading.

Wavelet transform has been successfully used [10–12] to analyze nonstationary signals in electrical machines to evaluate faults in transient state. Wavelet-based technique through successive adaptive signal cancelation utilizing HPF and LPF results in producing wavelet coefficient which distinguishes between healthy and faulty motor. Scaling of the amplitude and shifting of the frequency make the analysis more effective for transient or nonstationary signals.

In the present work, the starting current has been analyzed using DWT. The decomposed detailed coefficients of the starting current of the faulty motor have been compared to those of the healthy motor. Statistical coefficients viz. variation between maximum and minimum values, rms value, mean value, and standard deviation (STD) are used as fault-identifying parameters.

B.  Results and discussions

Using Example 5.1, transient current envelope has been captured with a sampling rate of 5120 sample/s., whereas supply frequency is 50 Hz having time period of 20 ms. The starting current envelops of healthy and mass unbalance rotor motors are shown in Figs. 5.19 and 5.20, respectively.

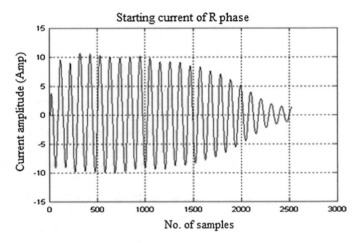

**Fig. 5.19**  Starting current of healthy motor

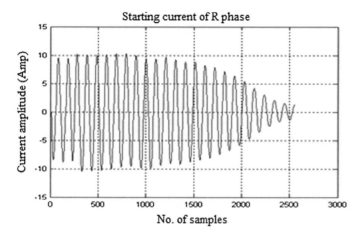

**Fig. 5.20**   Starting current of mass unbalance rotor motor

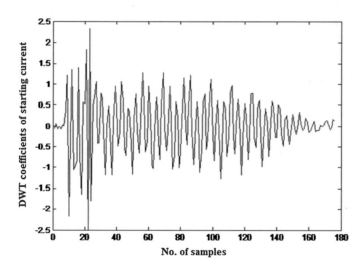

**Fig. 5.21**   DWT plot at level 4 for healthy motor

The wavelet decomposition of the starting current signals is performed for both the motors. Maximum variations of the detailed coefficients of the DWT of the signals are observed in the decomposition levels 4th, 5th, and 6th, and these are shown in Figs. 5.21 and 5.22 for 4th level, in Figs. 5.23 and 5.24 for 5th level, and Figs. 5.25 and 5.26 for 6th level for healthy and mass unbalanced rotor motor, respectively. The statistical parameters of the detailed coefficients are extracted. It is observed that the variations are distinct in the above three levels. The variations are much more prominent in the 5th level and are much higher in case of faulty motor than those for healthy motor. The variations are also higher for 4th level but not as

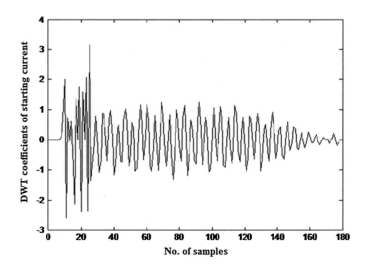

**Fig. 5.22** DWT plot at level 4 for mass unbalance rotor motor

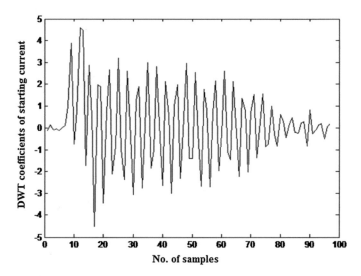

**Fig. 5.23** DWT plot at level 5 for healthy motor

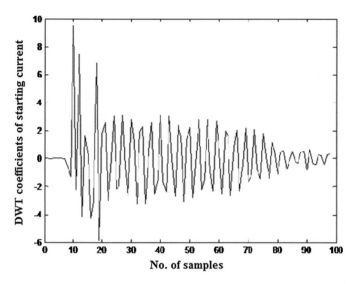

**Fig. 5.24** DWT plot at level 5 for mass unbalance rotor motor

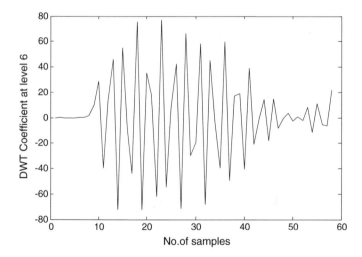

**Fig. 5.25** DWT plot at level 6 for healthy motor

much as in the 5th level. The 6th level also gives higher value of statistical parameters for faulty motor than those for the healthy motor except the value of (max–min) and mean which are less for faulty motor, as shown in Table 5.7.

So it may be inferred that the 5th level decomposition is most suitable for detection of this mass unbalance rotor fault of an induction motor.

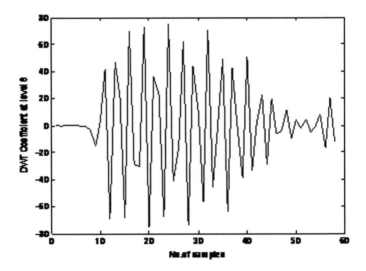

**Fig. 5.26** DWT plot at level 6 for mass unbalance rotor motor

## 5.11   Diagnosis by PDD of Reconstructed Starting Current After Wavelet Transform [13]

A.  Theoretical development

The wavelet transform is given by (5.6)

$$C(a,b) = \frac{1}{\sqrt{a}} \int_{-\infty}^{\infty} x(t)\Psi\left(\frac{t-b}{a}\right) dt \qquad (5.6)$$

where $x(t)$ is the signal, $a$ and b are the real quantities and denote the wavelet scale and position, and $\psi(t)$ is the wavelet. There are two types of wavelet transforms: continuous (CWT) and discrete (DWT). Although CWT gives more accurate result, DWT is more beneficial in practical application because of utilizing discrete range and shorter computational time.

The spectral frequency bands at different frequency levels are shown in Table 5.8 where the sampling frequency for the present experiment was 5120 cycles/s.

In this attempt, DWT is applied on the starting current envelop of the induction motors. The family of Daubechies wavelets is chosen as the basis function for the analysis. To avoid ambiguous diagnosis, Daubechies wavelets of the order of 10, i.e., "db10," which have higher smoothness [14], have been used.

**Table 5.8** Spectral frequency bands at different decomposition levels

| Decomposition detail | Frequency band |
|---|---|
| Detail at level 1 | 1280–2560 |
| Detail at level 2 | 640–1280 |
| Detail at level 3 | 320–640 |
| Detail at level 4 | 160–320 |
| Detail at level 5 | 80–160 |
| Detail at level 6 | 40–80 |
| Detail at level 7 | 20–40 |
| Detail at level 8 | 10–20 |
| Approximate at level 8 | 0–10 |

**Table 5.9** Comparison of decay time of transient current

| Motor condition | Figure no. | Time to reach steady state (ms) |
|---|---|---|
| Healthy | Figure 5.21 | 494.14 |
| Mass unbalanced rotor | Figure 5.22 | 501.95 |

Starting current being less sensitive to load and having large value, about 8–10 times the rated current, the starting transient current at no load has been considered to avoid running the motor at full load, in this work. Also, an improved detection technique of mass unbalance in rotor has been developed using DWT. The technique is the use of a parameter, called PDD, defined as square of DWT coefficients [15].

B. Results and discussions

The starting current envelops under no load as shown in Figs. 5.19 and 5.20 reveal minor difference between healthy and faulty (mass unbalance in rotor) motors. In faulty motor, the decay time of the transient current is more compared to healthy motor, i.e., faulty motors take longer time to reach at steady-state condition and this is depicted in Table 5.9. The sampling frequency rate in capturing these current envelops is 5120 samples/s. The captured signals after decomposition into detailed and approximate coefficient up to level 8 are reconstructed. Figures 5.27 and 5.28 show the PDD plot of 4th, 5th, and 6th levels of reconstructed details of healthy and faulty motors, respectively, in which the differences between the healthy and faulty motors are distinctly visible and this is shown in Table 5.10. The mean values of the PDD parameters are much more for faulty motor, especially for 6th level which distinguishes the faulty motor from the healthy motor.

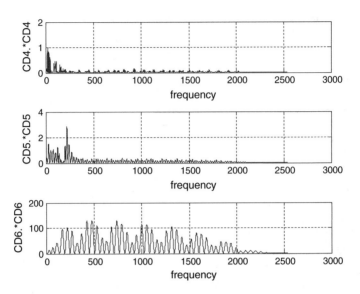

**Fig. 5.27** PDD plot of 4th, 5th, and 6th levels of reconstructed details of healthy motor

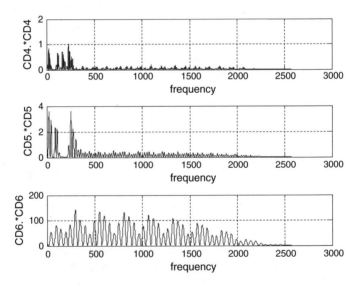

**Fig. 5.28** PDD plot of 4th, 5th, and 6th levels of reconstructed details of motor with mass unbalance rotor

**Table 5.10** Comparison of PDD values of detailed coefficients

| Motor condition | Figure no. | Mean value of PDD at wavelet level | | |
|---|---|---|---|---|
| | | 4 | 5 | 6 |
| Healthy | Figure 4.29 | 0.03355 | 0.1443 | 29.88 |
| Mass unbalanced rotor | Figure 4.30 | 0.04405 | 0.2017 | 32.14 |

## 5.12  Conclusions

Different techniques to diagnose mass unbalance fault of an induction motor have been described. First, the conventional method, FFT, is performed on the steady-state motor vibration signal and current signal but the result is too prominent to differentiate between healthy and faulty motors. In the second technique, radar analysis of stator current Concordia is performed on starting current. Here, first Concordia of starting current is formed, and then by developing radar diagrams of these Concordia, analysis is performed. Different lengths of average diameters of radar diagrams (which are nearly circular in shape) confirm the presence of fault. In the third technique, DWT is performed on the transient starting current of the motor. Then statistical parameters like root mean square (RMS), standard deviation (STD), and mean values of DWT coefficients at different decomposition levels are calculated. From the result, it is concluded that the 5th decomposition level is the most suitable for detection of the rotor mass unbalance fault. In the last technique shown, mean value of PDD is calculated from the DWT coefficients at different decomposition levels. It is observed that at 5th and 6th levels these values are much more for faulty motors compared to those of healthy motor.

## References

1. Karmakar S, Ahamed SK, Mitra M, Sengupta S (2007) Diagnosis of fault due to unbalanced rotor of an induction motor by analysis of vibration and motor current signatures. International conference MS'07, India, 3–5 Dec 2007, pp 399–403
2. Cusido J, Rosero J, Aldabas E, Ortega JA, Romeral L (2006) New fault detection techniques for induction motors. Electr Power Quality Utilisation, Magazine II(1)
3. Benbouzid MEH (2000) A review of induction motor signature analysis as a medium for fault detection. IEEE Trans Ind Electron 47(5):984–993
4. Chattopadhyay S, Karmakar S, Mitra M, Sengupta S (2012) Radar analysis of stator current Concordia for diagnosis of unbalance in mass and cracks in rotor bar of an squirrel cage induction motor. Int J Model Meas Control Gen Phys Electr Appl, AMSE, Series A, 85(1):50–61. ISSN: 1259-5985
5. Chattopadhyay S, Mitra M, Sengupta S (2007) Harmonic analysis in a three-phase system using park transformation technique. Modeling-A. AMSE Int J Model Simul, France, 80 (3):42–58
6. Chattopadhyay S, Mitra M, Sengupta S (2011) Electric power quality. First edn, Springer
7. Ahamed SK, Karmakar S, Mitra M, Sengupta S (2009) Detection of mass unbalance rotor of an induction motor using wavelet transform of the motor starting current at no load. In: Proceedings of national conference on modern trends in electrical engineering (NCMTEE-2009), organized by IET and HETC, Hooghly, West Bengal, pp MC-1 to MC-6, 11–12 July 2009
8. Ho Ha K, Hong JP, Kim G-T, Chang K-C, Lee J (2000) Orbital analysis of rotor due to electromagnetic force for switch reluctance motor. IEEE Trans on Magnetics 36(4)
9. Nandi S, Toliyat HA (1999) Condition monitoring and fault diagnosis of electrical machines—a review. In: Proceedings 34th annual meeting of IEEE industrial applications society, pp 197–204

10. Douglas H, Pillay P, Ziarani AK (2004) A new algorithm for transient motor current signature analysis using wavelets. IEEE Trans Ind Appl 40(5):1361–1368
11. Chow TW, Hai S (2004) Induction machine fault diagnostic analysis with wavelet technique. IEEE Trans Ind Electron 51(3)
12. Wilkinson WA, Cox MD (1996) Discrete wavelet analysis of power system transients. IEEE Trans Power Syst 11(4)
13. Ahamed SK, Karmakar S, Mitra M, Sengupta S (2010) Novel diagnosis technique of mass unbalance in rotor of induction motor by the analysis of motor starting current at no load through wavelet transform. In: 6th International conference on electrical and computer engineering, ICECE 2010, 18–20 Dec 2010, Dhaka, Bangladesh, pp 474–477, 978-1-4244-6279-7/10©2010 IEEE IEEE Xplore
14. Douglas H, Pillay P (2005) The impact of wavelet selection on transient motor current signature analysis. 0-7803-8987-5/05 ©2005, IEEE
15. Cusido J, Jornet A, Romeral L, Ortega JA, Garcia A (2006) Wavelet and PSD as a fault detection technique. ITMC 2006-instrumentation and measurement technology conference. Sorrento, Italy. 24–27 Apr 2006

# Chapter 6
# Stator Winding Fault

**Abstract** The chapter deals with stator winding faults. Assessments of different types of stator winding fault are discussed in this chapter. Both steady-state and transient current-based diagnosis are covered. Sequence component-based stator fault diagnosis is presented. Park plane is sometime used for fault diagnosis. Different wavelet transformation techniques are used for stator fault diagnosis.

**Keywords** DWT · FFT · Negative sequence component · Park plane · Statistical parameters · Stator winding fault · Transients

**Chapter Outcome**

After completion of the chapter, readers will be able to gather knowledge and information regarding following areas:

- Stator winding fault
- Steady-state approaches for broken bar assessment
- Diagnosis using starting transients
- Sequence component-based assessment
- Assessment in Park plane
- Diagnosis using wavelet transforms.

## 6.1 Introduction

This chapter describes methods for assessment of stator winding fault. First, the fault in induction motor is discussed in general then the fault is diagnosed by motor current signature analysis (MCSA) technique using different signal processing tools. Finally, a comparison among different outcomes, their advantages, and disadvantages are shown in the conclusion of the chapter.

© Springer Science+Business Media Singapore 2016
S. Karmakar et al., *Induction Motor Fault Diagnosis*,
Power Systems, DOI 10.1007/978-981-10-0624-1_6

## 6.2   Stator Winding Fault

When failure of insulation of the stator winding of an induction motor occurs then it is referred as the stator winding fault. It is a very common stator fault of an induction motor. Different types of stator winding faults are (i) short circuit between two turns of same phase—called turn-to-turn fault, (ii) short circuit between two coils of same phase—called coil to coil fault, (iii) short circuit between turns of two phases—called phase to phase fault, (iv) short circuit between turns of all three phases, (v) short circuit between winding conductors and the stator core—called coil to ground fault, and (vi) open circuit fault when winding gets disconnected. As per the review of IEEE and EPRI, 28–36 % of induction motor faults are stator winding fault [1, 2].

Causes of stator winding faults are different stresses namely mechanical stress, electrical stress, thermal stress, and environmental stress. Mechanical stress happens mainly due to movement of stator coil. Force due to large value of stator current (as force is proportional to the square of current [3]) produces movement of stator coil. Also high mechanical vibration of the motor may cause mechanical stress. Electrical stress happens when transients are present in the supply voltage. Thermal overloading is the main reason of thermal stress which deteriorates the insulation of the stator winding. Thermal overloading occurs mainly due to over-current drawn by the motor or unbalanced supply voltage or improper ventilation or higher ambient temperature etc. [4]. The thumb rule states that for every 10 °C rise in ambient temperature, the life of insulation decreases by 60 % [5]. Environmental stress may occur on the motor if it is operated in a hostile environment like too hot or cold or too humid atmosphere.

Different methods to diagnose stator winding fault are shown in the following sections.

## 6.3   Useful Analytical Tools for Stator Winding Fault Diagnosis

For diagnosing stator winding fault, both transient and steady stator currents have been captured and are analyzed using different signal processing tools as shown in Fig. 6.1. At first, the diagnosis is performed using steady-state stator current by different conventional processes namely negative sequence component study, Park plane analysis and FFT analysis. Then diagnosis is performed using both steady-state and transient-state stator currents of the induction motor. In the last two methods, as shown in Fig. 6.1, starting transient current is analyzed using Wavelet Transform. All the methods are described in subsequent sections.

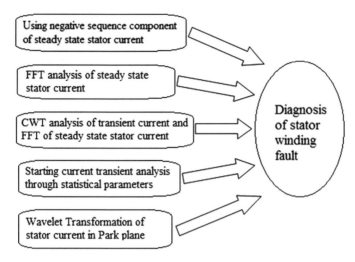

**Fig. 6.1** Diagnosis of stator winding fault using different techniques

## 6.4 Diagnosis of Inter-turn Short Circuit Fault Using Negative Sequence Component of Stator Current at Steady State [5]

During inter-turn short circuit condition small unbalance is created in stator current and it results in generation of negative sequence current. During unbalance, the three-phase currents and voltages consist of positive, negative, and zero sequence components. In matrix form, the phase currents can be written as [6]

$$
\begin{bmatrix} I_R \\ I_Y \\ I_B \end{bmatrix} = \begin{bmatrix} 1 & 1 & 1 \\ 1 & a^2 & a \\ 1 & a & a^2 \end{bmatrix} \begin{bmatrix} I_{R0} \\ I_{R1} \\ I_{R2} \end{bmatrix} \tag{6.1}
$$

$$
\begin{bmatrix} I_{R0} \\ I_{R1} \\ I_{R2} \end{bmatrix} = \frac{1}{3} \begin{bmatrix} 1 & 1 & 1 \\ 1 & a & a^2 \\ 1 & a^2 & a \end{bmatrix} \begin{bmatrix} I_R \\ I_Y \\ I_B \end{bmatrix} \tag{6.2}
$$

Thus, Eq. (6.2) gives the mathematical relation between resultant stator currents and sequence components present in the stator currents. Sequence components consist of zero, positive and negative sequences. Among them change of negative sequence component is significant. In the following section, experiment is carried out to assess negative sequence component generated due to inter turn short fault.

*Example 6.1* Experimentation is done for detection of negative sequence components as shown in Fig. 6.2. The induction motor, on which experiment is performed, is shown in Fig. 6.3 and its specification is given in Table 6.1. The stator winding of

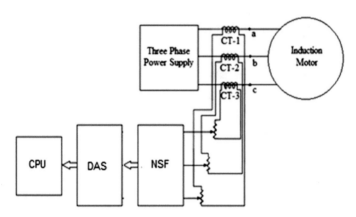

**Fig. 6.2** Experimental set up

**Fig. 6.3** Induction motor, used for the experiment (*courtesy* Power System Lab, Department of Applied Physics, University of Calcutta, India)

**Table 6.1** Specification of the motor used for the experiment

| Make | Power | Phase | Voltage | Frequency | Speed |
|---|---|---|---|---|---|
| Local (Kolkata) | 1 HP | 3 phase | 416 V | 50 Hz | 1460 rpm |

the motor is specially designed and having many tapping in all phases. The motor is started through direct on line (DOL) starter and starting current of all three phases are stepped down and then passed through negative sequence filter (NSF). Output of the filter is sampled and captured through a data acquisition system (DAS) and then sent to Central Processing Unit (CPU) of a computer. The sampling frequency of the DAS is 2048 samples/s. First, the motor is run at normal condition, i.e., there is no inter-turn sort circuit in stator. Then, by shorting the tapping, inter-turn short circuit is implemented at different percentage of the winding. At those conditions, again stator currents are stepped down, sampled, and then captured for assessment.

Details about fault detection are discussed in these sections of the chapter with the help of the results and observation obtained by experimentation with this Example 6.1.

**Results and discussions** Different inter-turn short conditions are created by shorting tapings of stator winding. Thus the motor is run at different inter-turn short conditions. Stator currents are collected through three current transformers. After stepping down, stator currents are passed through negative sequence filter. By this way, negative sequence components are extracted and then, sampled, digitized, and captured through data acquisition system. Negative sequence components present in stator current are measured at different percentage of inter turn short. Negative sequence component versus percentage of inter turn short is plotted in Fig. 6.4. It shows significant rise of negative sequence component with the increase of percentage of short. For small percentage of inter-turn short, the relationship is almost linear.

**Algorithm**—For unknown motor having same type of design features, it is possible, following the curve shown in Fig. 6.4, to predict the percentage of inter-turn short circuit in the stator winding of the motor. An algorithm for fault detection can be made as follows:

(a)  Step-down stator currents at starting using current transformer
(b)  Extract negative sequence component through negative sequence filter
(c)  Sample and capture of the negative sequence component through data acquisition system
(d)  Measure of the negative sequence component
(e)  Assess percentage of inter turn short in stator winding.

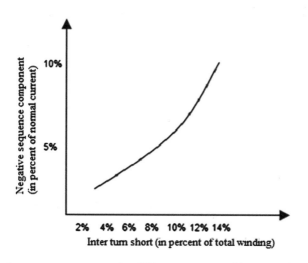

**Fig. 6.4**  Negative sequence component for different percentage of inter turn short

## 6.5   FFT-Based Inter-turn Short Circuit Analysis of Steady-State Stator Current [7]

Steady-state current: Steady-state current is a useful signal for motor fault diagnosis as it is available for long period of time which makes data capturing easier than transient period during switching on or off. In this section, FFT-based fault diagnosis has been done using steady-state motor current.

Due to inter-turn short, the changes in nature of the starting current transients are observed in this work and the variation of contour of wavelet coefficients is assessed for different percentage of inter-turn short of the stator winding. Also, after filtering the supply frequency from the stator current, FFT has been performed on the steady-state portion of the current.

Starting Current Transients: At normal operating condition, induction motor draws balanced current from the supply. Practically, in addition to the AC transient stator current also consist of small amount of DC during the starting of the induction motor. Mathematically,

$$i(t) = \Sigma_n i_n + i_{dc} = \Sigma_n I_n \sin(2\pi f_n t + \phi_n) + i_{dc}) \qquad (6.3)$$

where is the amplitude of harmonic component of frequency $f_n$, $\Phi_n$ is the corresponding phase angle, and $i_{dc}$ is the DC component present in the starting current. The harmonics present in the transients depend on the design parameters and operating condition of the motor. As the design parameters of the stator winding change due to inter-turn fault in stator winding, it is obvious, harmonic content and frequencies present in the stator current will also change.

Steady-state signals are assessed by FFT in this section using Example 6.2 whereas transients are assessed by CWT and DWT in next sections.

*Example 6.2* Three-phase induction motor, as given in Table 6.1, is used for experimentation. The stator winding of the motor is specially designed with many tapings in all three phases so that by shorting the tapings from outside, inter-turn short circuit fault can be implemented in the motor, at different percentage of the stator winding. The motor is started through direct on line (DOL) starter and stator currents are stepped down, sampled, and captured through a data acquisition system (DAS). The sampling frequency considered is 2048 samples/sec.

The schematic diagram is shown in Fig. 6.6. 3 ph, 340 V, 50 Hz, supply is provided to the designated induction motor. The experiments are performed under healthy condition of the motor and then with motor having the stator winding short circuit fault at (i) 2.5 % turn, (ii) 5 % turn, and (iii) 10 % turn of the total winding (Fig. 6.5).

Details about fault detection are discussed in these sections of the chapter with the help of results and observation obtained by experimentation with this Example 6.2.

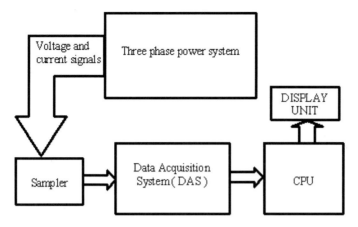

**Fig. 6.5** Schematic diagram of experimental set up

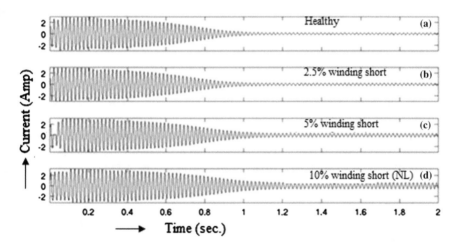

**Fig. 6.6** Stator current under no-load condition of the motor; **a** for healthy motor, **b** for motor when 2.5 % of stator winding is shorted, **c** for motor when 5 % turn of stator winding is shorted, and **d** for motor when 10 % turn of stator winding is shorted. Shorting is in one of the three phases

The three-phase stator current which is captured by Hall probe is shown in Fig. 6.6 and 6.7. In Fig. 6.6a, current shown is the no-load current when there is no fault in the motor. Then the fault is created by shorting the tapings at 2.5 % turns, 5 % turns, and 10 % turns of the stator windings and currents are captured which are shown in Fig. 6.6b, c, and d, respectively. Similarly in Fig. 6.7a, b, c, and d currents shown are under the loaded condition of the motor. It is observed that with load and increase of percentage of fault, time duration of the starting current transient increases considerably.

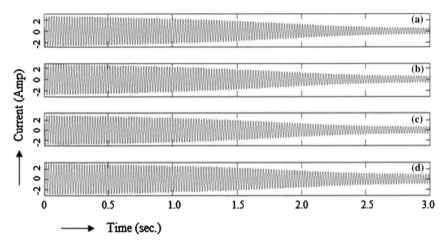

**Fig. 6.7** Stator current under load condition of the motor; **a** for healthy motor, **b** for motor when 2.5 % of stator winding is shorted, **c** for motor when 5 % turn of stator winding is shorted, and **d** for motor when 10 % turn of stator winding is shorted. Shorting is in one of the three phases

## 6.5.1 FFT-Based Assessment

To analyze the characteristic of the current of the healthy and faulty motor, in the present work the current signals are divided into two portions: (a) transient and (b) steady portion. FFT has been performed on the steady portion of the current signal to analyze the effect of faults on different harmonic and sidebands. Steady portion of stator currents are first filtered to band off supply frequencies (49–61) Hz then FFT has been performed. Result is shown in Figs. 6.8 and 6.9.

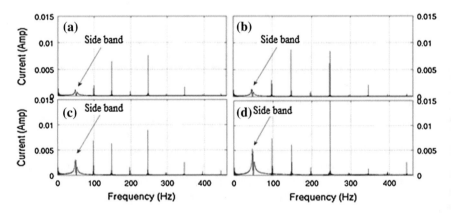

**Fig. 6.8** FFT of the steady state current signature of the motor under no-load condition; **a** for healthy motor, **b** for motor when 2.5 % turn of stator winding is shorted, **c** for motor when 5 % turn of stator winding is shorted, and **d** for motor when 10 % turn of stator winding is shorted

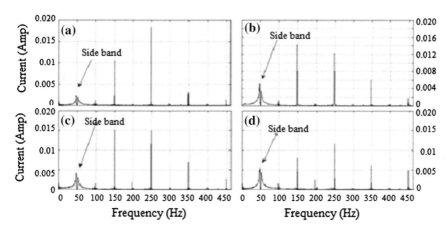

**Fig. 6.9** FFT of the steady-state current signature of the motor under load condition; **a** for healthy motor, **b** for motor when 2.5 % turn of stator winding is shorted, **c** for motor when 5 % turn of stator winding is shorted, and **d** for motor when 10 % turn of stator winding is shorted

## 6.5.2    Observations

It is observed from Fig. 6.8 that under no-load condition, when the fault occurs in the motor, magnitude of all the third, fifth, and seventh harmonics are increased compared to the healthy motor and under load condition (Fig. 6.9), magnitudes of the third and seventh harmonics are increased but the fifth harmonic decreases when compared to the healthy motor. The magnitude of sidebands has increased considerably for both load and no-load conditions. A comparison chart showing the change of magnitudes of sidebands of supply frequency is shown in Tables 6.2 and 6.3. Table 6.2 shows the comparison under no-load condition of the motor and Table 6.3 under load condition.

**Table 6.2** Percentage (%) increase in amplitude of sidebands of faulty motor at no-load

| Sidebands | % Increase in amplitude under no-load condition of motor with stator winding fault having | | |
|---|---|---|---|
| | 2.5 % Turn short | 5 % Turn short | 10 % Turn short |
| Left sideband of 50 Hz (%) | 18.6 | 171.8 | 289 |
| Right sideband of 50 Hz (%) | 8.7 | 122 | 268 |

**Table 6.3** Percentage (%) increase in amplitude of sidebands of faulty motor under load

| Sidebands | % Increase in amplitude under load condition of motor with stator winding fault having | | |
|---|---|---|---|
| | 2.5 % Turn short | 5 % Turn short | 10 % Turn short |
| Left sideband of 50 Hz (%) | 137.6 | 88.37 | 147.7 |
| Right sideband of 50 Hz (%) | 81.6 | 62.4 | 121.97 |

**Table 6.4** Total harmonic distortion (THD) based on FFT

|              | 0 % Short | 2.5 % Short | 5 % Short | 10 % Short |
|--------------|-----------|-------------|-----------|------------|
| THD (No-load)| 8.7766    | 9.473       | 7.3163    | 4.0727     |
| THD (Load)   | 7.171     | 8.3948      | 6.7118    | 2.9404     |

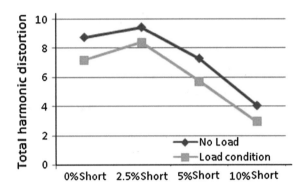

**Fig. 6.10** Total harmonic distortion versus percentage of inter turn short

Total harmonic distortion (THD) is shown in Table 6.4 and presented in Fig. 6.10. It shows significant variation of THD with respect to the percentage short of stator winding.

### 6.5.3   Inference

Tables 6.2 and 6.3 are obtained from Figs. 6.8 and 6.9 by measurement and calculations. Tables show percentage (%) increase in amplitude of left sidebands and right sidebands of 50 Hz for different percentage of inter-turn short fault of stator winding under no-load and load conditions, respectively. They reveal that both the left sideband and right sideband of 50 Hz increases with the increase of percentage of stator winding fault with respect to the healthy motor. Table 6.4 shows the value of total harmonic distortion (THD) for different percentage of inter-turn short fault of stator winding under no-load and load conditions. It is observed that THD decreases with the increase of percentage of stator winding fault under both no-load and load conditions.

## 6.6   CWT- and DWT-Based Inter-turn Short Circuit Assessment Using Transient Current [8]

Steady-state motor current depends largely on load condition, but starting current is less dependent on load condition. Nature of starting current largely depends mainly on motor parameters. Thus starting current is a very useful signal for motor fault diagnosis. In this section, continuous wavelet transform (CWT)- and discrete wavelet transform (DWT)-based fault assessments have been done using starting current at no-load condition.

### 6.6.1  Wavelet Transform

The wavelet transform is governed by (6.4)

$$C(a,b) = \frac{1}{\sqrt{a}} \int_{-\infty}^{\infty} x(t) \Psi\left(\frac{t-b}{a}\right) dt \qquad (6.4)$$

where $x(t)$ is the signal, a and b being real denote the wavelet scale and position, and $\psi$ is the wavelet function.

### 6.6.2  Discrete Wavelet Transform (DWT)

In this present work, the sampling frequency considered being 2048 samples/s, the frequency bandwidth of approximations and details are shown in Table 6.5.

### 6.6.3  Energy Calculation

Energy of the nonstationary starting current is calculated using Parseval's theorem. This theorem refers that the sum of the square of a function is equal to the sum of the square of its transform. Using wavelet coefficients Parseval's theorem can be stated as "the energy that a time domain function contains is equal to the sum of all

**Table 6.5** Frequency bandwidths at different wavelet levels

| Wavelet level | Approximations | Bandwidth (Hz) | Details | Bandwidth (Hz) |
|---|---|---|---|---|
| 1 | $a_1$ | 0–612 | $d_1$ | 612–1024 |
| 2 | $a_2$ | 0–266 | $d_2$ | 266–612 |
| 3 | $a_3$ | 0–128 | $d_3$ | 128–266 |
| 4 | $a_4$ | 0–64 | $d_4$ | 64–128 |
| 6 | $a_6$ | 0–32 | $d_6$ | 32–64 |
| 6 | $a_6$ | 0–16 | $d_6$ | 16–32 |
| 7 | $a_7$ | 0–8 | $d_7$ | 8–16 |
| 8 | $a_8$ | 0–4 | $d_8$ | 4–8 |

energy concentrated in the different decomposition levels of the corresponding wavelet transformed signal". This can be mathematically expressed as (6.6) as is mentioned in [9].

$$\sum_{n=1}^{N} |x(n)|^2 = \sum_{n=1}^{N} |a_j(n)|^2 + \sum_{n=1}^{m} \sum_{n=1}^{N} |d_j(n)|^2 \qquad (6.5)$$

where $x(n)$ is time domain discrete signal, N is total number of samples in the signal, $\sum_{n=1}^{N} |x(n)|^2$ is total wave energy of the signal $x(n)$, $\sum_{n=1}^{N} |a_j(n)|^2$ is total energy concentrated in the "$j$" wavelet level of the approximated version of the signal, $\sum_{n=1}^{m} \sum_{n=1}^{N} |d_j(n)|^2$ is total energy concentrated in the detail version of the signal, from level 1 to $m$, $m$ is maximum level of wavelet decomposition, $a_j$ is the approximate coefficient, and $d_j$ is the detail coefficient of $j$th wavelet level.

*Example 6.2* One test rig manufactured by M/S Spectra Quest, USA is used to carry out the experiment. Experimental set up is shown in Fig. 6.11. Specially designed three-phase 440 V, 50 Hz, 1 HP motor, as shown in Fig. 6.3, is used. The stator winding of the motor is specially designed and having many tapings in all the three phases to implement short-circuited stator winding fault from outside. The motor is started through direct on line (DOL) methods and current is captured by a Hall probe through a data acquisition system (DAS). Specifications of Hall probe and DAS are shown in Table 6.6. The sampling frequency considered is 2048 Hz. At first, the motor is run at normal condition, i.e., there being no inter-turn short circuit in the stator. Then, by shorting the tapping, inter-turn short circuit are implemented at (i) 2.5 % turn, (ii) 5 % turn, and (iii) 10 % turn of total winding in

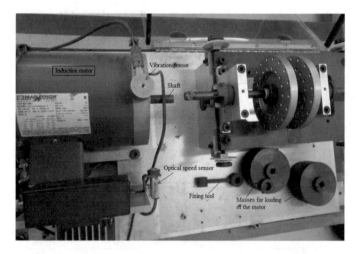

**Fig. 6.11** Experimental set up (*courtesy* Power System Lab, Department of Applied Physics, University of Calcutta, India)

**Table 6.6** Specifications of hall probe and data acquisition system

| Hall probe | | | Data acquisition system | | |
|---|---|---|---|---|---|
| Make | Model | Conversion | Make | Model | Speed |
| LEM | PR30 ACV 600 V CATIII | 30A ac/3 V ac | OROS | OR36, 8 channels | 100 mbps |

one phase of the stator. At these conditions, again stator currents are captured for assessment. 3ph, 340 V, 50 Hz, supply is provided to the designated induction motor.

Details about fault detection are discussed in the subsequent sections of the chapter with the help of results and observation obtained by experimentation with this Example 6.2.

### 6.6.4 Results and Discussions

The captured stator currents are shown in Figs. 6.6 and 6.7 under no-load and load condition of the motor, respectively. The transient part of the captured motor current signature is analyzed by CWT and the steady part is first filtered to band off the supply frequency and then with this filtered signal (i) FFT-based analysis has been performed and also (ii) energy has been calculated after performing the DWT.

#### 6.6.4.1 CWT- Based Transient Analysis

Transient part of the current signature is assessed using CWT with 'db10' wavelet. Contour of the wavelet coefficients are plotted as shown in the Figs. 6.12 and 6.13 for no-load and load condition, respectively indicating percentage energy for each coefficient.

Low-scale values correlate with high frequency content of the stator current signature and high-scale values correlate with low frequency content of the stator current signature. Figure 6.14 represents the relation between frequencies and scale of CWT. From Figs. 6.12, 6.13, and 6.14 it is observed that in the transient portion, frequencies those are generated in the region of 30 and 70 Hz have very small amplitude ($0.6 \times 10^{-3}$) whereas frequencies generated in the region of 38 and 48 Hz have higher amplitude (approx. $2.6 \times 10^{-3}$).

#### 6.6.4.2 DWT-Based Steady State Analysis by Energy Calculation

The steady-state current of the healthy motor and stator winding faulty (turn-to-turn fault) motor are filtered to band off supply frequency. Then DWT is performed on

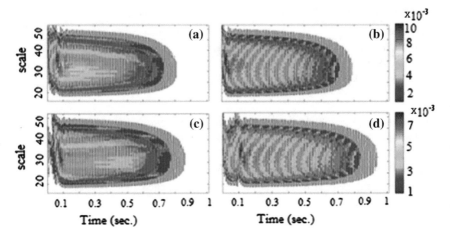

**Fig. 6.12** Contour of the CWT coefficients of transient current signature of the motor under no-load; **a** for healthy motor, **b** for motor when 2.5 % turn of stator winding is shorted, **c** for motor when 5 % turn of stator winding is shorted, and **d** for motor when 10 % turn of stator winding is shorted

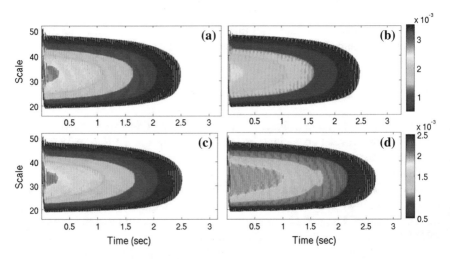

**Fig. 6.13** Contour of the CWT coefficients of transient current signature of the motor under load; **a** for healthy motor, **b** for motor when 2.5 % turn of stator winding is shorted, **c** for motor when 5 % turn of stator winding is shorted and **d** for motor when 10 % turn of stator winding is shorted

these filtered signals, using mother wavelet 'db10', up to wavelet level 8 to find the approximate coefficients ($a_j$) and detail coefficients ($d_j$). Using these $a_j$ and $d_j$ approximated wave energy and detailed wave energy are calculated from the right hand side of (6.6) for each level. Detail wave energies under no-load and load

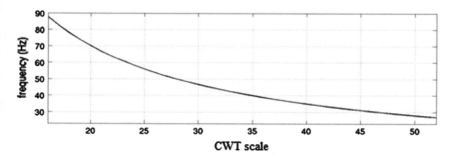

**Fig. 6.14** Conversion of CWT scales to frequency

**Table 6.7** Detail wave energy at different wavelet level under no-load

| Wavelet level | Healthy no-load | 2.5 % Short no-load | 5 % Short no-load | 10 % Short no-load |
|---|---|---|---|---|
| 1 | 0.0102 | 0.016 | 0.0263 | 0.098 |
| 2 | 0.4746 | 0.6662 | 0.6662 | 1.2699 |
| 3 | 1.8118 | 2.242 | 1.7161 | 3.3667 |
| 4 | 2.097 | 2.8074 | 2.7083 | 4.6698 |
| 6 | 2.6437 | 3.8127 | 6.4199 | 10.846 |
| 6 | 2.6482 | 3.8234 | 6.4447 | 10.9686 |
| 7 | 2.6611 | 3.8326 | 6.4694 | 11.0902 |
| 8 | 2.6647 | 3.8433 | 6.6006 | 11.3124 |

**Table 6.8** Detail wave energy at different wavelet level under load

| Wavelet level | Healthy load | 2.5 % Short load | 5 % Short load | 10 % Short load |
|---|---|---|---|---|
| 1 | 0.0962 | 0.0676 | 0.1383 | 0.2307 |
| 2 | 2.9938 | 1.6403 | 2.03 | 2.3304 |
| 3 | 8.461 | 4.6913 | 6.0264 | 4.8169 |
| 4 | 9.339 | 6.3367 | 7.1609 | 6.7073 |
| 6 | 13.0343 | 10.2168 | 14.2931 | 19.2669 |
| 6 | 13.0864 | 10.3261 | 14.4096 | 19.4127 |
| 7 | 13.1186 | 10.4796 | 14.6964 | 19.6234 |
| 8 | 13.1676 | 10.8124 | 14.9004 | 19.6689 |

conditions at different levels are shown in Tables 6.7 and 6.8, respectively. In 6th, 6th and 7th decomposition level of detailed energy, distinct variations are observed between healthy and faulty motors and these are shown in Figs. 6.15 and 6.16 for motors under 'no-load' and 'load' conditions, respectively.

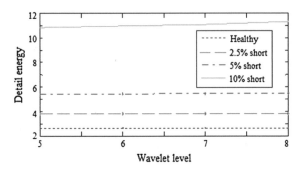

**Fig. 6.15**  Detail wave energy under no-load condition of motor

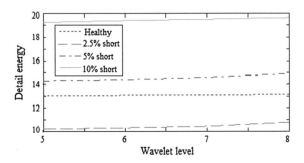

**Fig. 6.16**  Detail wave energy under load condition of motor

## 6.6.5  Inference

In this work, one of the various types of stator winding fault called turn-to-turn fault diagnosis of an induction motor is proposed by analyzing stator current both transient and steady-state parts. Techniques used for analyzing are Wavelet Transform (WT) and Fast Fourier Transform (FFT). On the transient part, Continuous Wavelet Transform (CWT) has been performed. Using CWT coefficients, contour has been formed which shows that in between CWT scale 30–40, amplitude is maximum and from the graph of conversion of CWT scale to frequency it is observed that this higher amplitude occurs in the frequency zone 38–48 Hz.

In the second part of the diagnosis, filtering out the supply frequency from the steady part of the stator current FFT has been performed. It is observed that under no-load condition of the motor when fault occurs, magnitudes of all the third, fifth, and seventh harmonics increase and under load condition magnitudes of the third and seventh harmonics increase but magnitude of fifth harmonic decreases in comparison to the healthy motor. The magnitude of sidebands has increased considerably for both load and no-load conditions. Also Discrete Wavelet Transform (DWT) has been performed on this filtered off steady-state current. Using DWT coefficients, detailed wave energy has been calculated by Parseval's theorem at different wavelet levels. Distinct variations are observed in the 6th, 6th and 7th level which corresponds to bandwidth 16–64 Hz.

From the results and observation of the work following conclusions can be made:

(i)   For turn-to-turn fault diagnosis, analysis of stator current can be carried out.
(ii)  The load condition of the motor and percentage of stator winding short affect the time duration of the transient portion of the stator current considerably.
(iii) Magnitudes of 3rd and 7th harmonic increase in the steady state of the stator current.
(iv)  Amplitude of the lower frequencies (below supply frequency, 50 Hz) becomes prominent.
(v)   In this proposed work, both WT and FFT are used. FFT can extract information from steady portion of current whereas WT can extract information both from transient and steady portion of current effectively.

## 6.7  Stator Winding Fault Diagnosis ByStarting Current Transient Analysis Through Statistical Parameters [10]

### 6.7.1  Skewness and Kurtosis

Skewness is a measure of asymmetry of data. A data set is said to be asymmetry if it does not look same to the left and right of the center point. Skewness of a normal distribution is zero. A data set will be as symmetric as its skewness is near to zero. If skewness is negative, distribution of data is concentrated on the right of the center point and the distribution is said to be left skewed. Similarly, positive skewness means distribution of data is concentrated on the left of center point and is called right skewed. Skewness is expressed as

$$\text{Skewness} = \frac{\sum_{i=1}^{n} (x_i - \bar{x})^3}{(n-1)\sigma^3} \tag{6.6}$$

where $x_i$ is data, $n$ is number of data, $\bar{x}$ is mean of data and $\sigma$ is standard deviation.

Kurtosis is used to check whether data set are peaked or flat relative to a normal distribution. Data set have a peak or a flat top near the mean if they have high or low Kurtosis values, respectively. Kurtosis is defined as

$$\text{Kurtosis} = \frac{\sum_{i=1}^{n} (x_i - \bar{x})^4}{(n-1)\sigma^4} - 3 \tag{6.7}$$

where $x_i$ is data, $n$ is number of data, $\bar{x}$ is mean of data and $\sigma$ is standard deviation. This definition gives kurtosis of standard normal distribution as zero.

*Example 6.3* Three-phase 440 V, 50 Hz, 1 HP motor is used for experimentation as shown in Fig. 6.3. The stator winding of the motor is specially designed having many tapings (at 2.5, 5, 10, and 15 % of stator turns) in all phases. The motor is started through direct on line (DOL) starter and starting current from all phases are stepped down, sampled, and captured through a data acquisition system (DAS). Specifications of Hall probe and DAS are shown in Table 6.6. The sampling frequency is 2048 samples/sec. At first, the motor is run at normal condition, i.e., there is no inter-turn short circuit in stator—this is considered as healthy motor. Then, by shorting the tapping at 5 and 10 % respectively, inter turn short circuit of the stator winding is generated—in this condition it is considered as stator winding fault motor. Under these conditions also, stator currents are stepped down, sampled, and then captured for assessment. In both cases, motor is run under no-load and load conditions.

Details about fault detection are discussed in the subsequent sections of the chapter with the help of results and observation obtained by experimentation with this Example 6.3.

## 6.7.2   Results and Observation

The stator currents are assessed by Continuous Wavelet Transform (CWT). They are decomposed at different levels, using "db10" as mother wavelet. Then, different statistical parameters of the wavelet coefficients at different levels are determined. The Kurtosis value and Skewness of wavelet coefficients at different wavelet levels are shown in Figs. 6.17 and 6.18, respectively. Figure 6.17 shows very distinct Kurtosis level for each case. Here all lines are almost straight and horizontal. Kurtosis level is highest at healthy condition (i.e. for no inter-turn fault) and it decreases with the increase of percentage of inter-turn fault.

Figure 6.18 shows that skewness of CWT coefficient for healthy and faulty condition is straight and horizontal and it increases with load. Skewness of CWT coefficients decreases with the increase of percentage of inter-turn fault. Thus Kurtosis and skewness of wavelet coefficients can be used for diagnosis of inter-turn fault.

## 6.7.3   Algorithm

An algorithm for fault detection has been made as follows:

(a)  Step-down stator currents at starting
(b)  Sample current signal
(c)  Capture them through data acquisition system

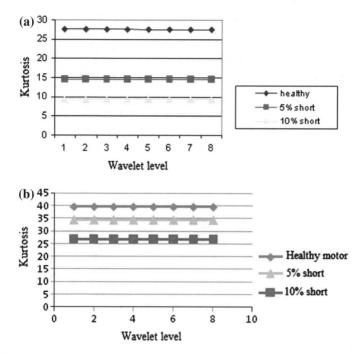

**Fig. 6.17** Kurtosis of CWT coefficient versus CWT levels of starting current transients of an induction motor **a** under no-load condition and **b** under load condition

(d)  Perform Continuous Wavelet Transformation of this starting current.
(e)  Determine of Kurtosis and Skewness of wavelet coefficients at different levels.
(f)  Detect the inter-turn fault based on the Kurtosis and Skewness of wavelet coefficients.

# 6.8  Wavelet Transformation-Based Stator Current Assessment in Park Plane [11]

## 6.8.1  Theoretical Development

*Impedance model of induction motor*: It depends on the value of the rotor resistance ($R_r$), stator resistance ($R_s$), rotor reactance ($X_r$), stator reactance ($X_s$), slip (s) and mutual inductance ($X_m$). The impedance model of an induction motor is given as follows [6]

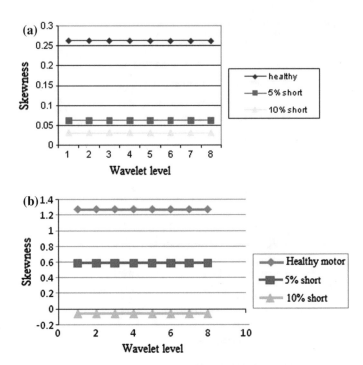

**Fig. 6.18** Skewness of CWT coefficient versus CWT levels of starting current transients of an induction motor **a** under no-load condition and **b** under load condition

$$
Z_m = \begin{pmatrix}
R_s + jX_s & jX_m & 0 & 0 \\
jX_m & R_r + jX_r & \omega X_r & \omega X_m \\
-\omega X_m & -\omega X_r & R_r + jX_r & jX_m \\
0 & 0 & jX_m & R_s + jX_s
\end{pmatrix}
\tag{6.8}
$$

*Computer simulation*: Mathematical modeling based on (6.9) is developed in MATLAB, considering stator reactance $X_s = \omega L_s$ and rotor reactance $X_r = \omega L_r$ with the following specifications and assuming both $d$-axis rotor voltage ($V_{dr}$) and $q$-axis rotor voltage ($V_{qr}$) as zero.

*Example 6.4* A motor is considered having following parameters:
Angular velocity ($\omega$) of the motor = 314 rad/s
slip of the motor = 0.04
rotor resistance ($R_r$) = 1.083 Ω
stator resistance ($R_s$) = 1.116 Ω
rotor inductance ($L_r$) = 0.006947 H
stator inductance ($L_s$) = 0.006947 H
mutual inductance ($X_m$) = 0.2037 H

Details about fault detection are discussed in the subsequent sections of the chapter with the help of results and observation obtained by experimentation with this Example 6.4.

Using Example 6.4 from Eq. (6.8)

$$Z_m = \begin{pmatrix} 1.115 + j1.875 & j0.203 & 0 & 0 \\ j0.203 & 1.083 + j1.87 & 1.12 & 0.12 \\ -0.12 & -1.12 & 1.083 + j1.87 & j0.2 \\ 0 & 0 & j0.2 & 1.115 + j1.875 \end{pmatrix} \quad (6.9)$$

*Transformation:* Three-phase balanced supply is fed to the motor for assessment. Three-phase stator voltage signals are transformed into direct and quadrature (*d–q*) axes by Park's matrix, given in (6.10). From these *d–q* voltages, currents in *d–q* plane are derived using the mathematical model, as given in (6.8) and (6.9). Using inverse Park's matrix, the currents in *d–q* plane are transformed into line currents which are used as signals for assessment of inter-turn short circuit fault of stator winding.

$$\begin{pmatrix} V_d \\ V_q \end{pmatrix} = \sqrt{\frac{2}{3}} \times \begin{pmatrix} \cos\theta & \cos(\theta - 2\pi/3) & \cos(\theta - 4\pi/3) \\ -\sin\theta & -\sin(\theta - 2\pi/3) & -\sin(\theta - 4\pi/3) \end{pmatrix} \times \begin{pmatrix} V_A \\ V_B \\ V_C \end{pmatrix}$$

$$(6.10)$$

### 6.8.2   RMS Values of Approximate and Detail Coefficients

Wavelet transform is performed on the line currents at different levels ($j$) to obtain detail ($d_j$, $j$ = 1 to 9) and approximate ($a_j$, $j$ = 1 to 9) coefficients. Root mean square (r.m.s.) of approximate coefficients ($a_j$) and detail coefficients ($d_j$) are determined at different percentage of stator winding short circuit (healthy condition i.e. 0 % winding short, 5 % winding short, 10 % winding short, 15 % winding short, 20 % winding short) conditions for all the three phases A, B, and C. All these results are presented in Tables 6.9, 6.10, 6.11, 6.12, 6.13, 6.14, 6.15, 6.16, 6.17, 6.18, 6.19, 6.20, 6.21, 6.22, and 6.23.

**Table 6.9**   Result for healthy motor, i.e., under normal condition of phase A

| Condition | Approximate coefficient | R.M.S. value | Detail coefficient | R.M.S. value |
|---|---|---|---|---|
| For current of phase A under normal condition | a1 | 0.240 + 0.062j | d1 | 0.0016 − 0.002j |
| | a2 | 0.241 + 0.061j | d2 | 0.009 + 0.016j |
| | a3 | 0.249 + 0.062j | d3 | 0.044 − 0.073j |
| | a4 | 0.271 + 0.001j | d4 | 0.096 + 169j |
| | a6 | 0.27 − 0.0001j | d6 | 0.0077 − 0.013j |
| | a6 | 0.27 + 0.0002j | d6 | 0.0086 + 0.014j |
| | a7 | 0.280 − 0.008j | d7 | 0.0074 − 0.012j |
| | a8 | 0.278 − 0.011j | d8 | 0.0046 + 0.007j |
| | a9 | 0.286 − 0.023j | d9 | 0.0083 − 0.013j |

**Table 6.10**   Result for 5 % winding short condition of phase A

| Condition | Approximate coefficient | R.M.S. value | Detail coefficient | R.M.S. value |
|---|---|---|---|---|
| For current of phase A under 5 % winding short | a1 | 0.242 + 0.067j | d1 | 0.0017 − 0.002j |
| | a2 | 0.243 + 0.066j | d2 | 0.0102 + 0.016j |
| | a3 | 0.262 + 0.068j | d3 | 0.048 − 0.0767j |
| | a4 | 0.272 + 0.00098j | d4 | 0.1013 + 0.163j |
| | a6 | 0.272 − 0.00018j | d6 | 0.0083 − 0.013j |
| | a6 | 0.274 + 0.0002j | d6 | 0.0092 + 0.014j |
| | a7 | 0.28 − 0.0089j | d7 | 0.008 − 0.012j |
| | a8 | 0.279 − 0.011j | d8 | 0.006 + 0.008j |
| | a9 | 0.287 − 0.024j | d9 | 0.0089 − 0.014j |

**Table 6.11**   Result for 10 % winding short condition of phase A

| Condition | Approximate coefficient | R.M.S. value | Detail coefficient | R.M.S. value |
|---|---|---|---|---|
| For current of phase A under 10 % winding short | a1 | 0.246 + 0.0623j | d1 | 0.00186 − 0.0027j |
| | a2 | 0.246 + 0.0612j | d2 | 0.011 + 0.0164j |
| | a3 | 0.266 + 0.074j | d3 | 0.0618 − 0.077j |
| | a4 | 0.272 + 0.00096j | d4 | 0.1118 + 0.166j |
| | a6 | 0.272 − 0.0002j | d6 | 0.009 − 0.0136j |
| | a6 | 0.274 + 0.00016j | d6 | 0.010 + 0.0162j |
| | a7 | 0.2814 − 0.0091j | d7 | 0.0087 − 0.013j |
| | a8 | 0.2797 − 0.0116j | d8 | 0.0064 + 0.0082j |
| | a9 | 0.2889 − 0.0248j | d9 | 0.0096 − 0.0144j |

**Table 6.12**  Result for 15 % winding short condition of phase A

| Condition | Approximate coefficient | R.M.S. value | Detail coefficient | R.M.S. value |
|---|---|---|---|---|
| For current of phase A under 15 % winding short | a1 | 0.2499 + 0.067j | d1 | 0.002 − 0.0028j |
| | a2 | 0.2603 + 0.066j | d2 | 0.0119 + 0.0168j |
| | a3 | 0.2696 + 0.081j | d3 | 0.0660 − 0.079j |
| | a4 | 0.272 + 0.0009j | d4 | 0.1208 + 0.170j |
| | a6 | 0.273 − 0.0002j | d6 | 0.0097 − 0.013j |
| | a6 | 0.276 + 0.0000j | d6 | 0.0108 + 0.0166j |
| | a7 | 0.282 − 0.009j | d7 | 0.0094 − 0.013j |
| | a8 | 0.280 − 0.0118j | d8 | 0.00684 + 0.008j |
| | a9 | 0.2902 − 0.0026j | d9 | 0.0104 − 0.014j |

**Table 6.13**  Result for 20 % winding short condition of phase A

| Condition | Approximate coefficient | R.M.S. value | Detail coefficient | R.M.S. value |
|---|---|---|---|---|
| For current of phase A under 20 % winding short | a1 | 0.266 + 0.0729j | d1 | 0.0021 − 0.0028j |
| | a2 | 0.266 + 0.071j | d2 | 0.0128 + 0.017j |
| | a3 | 0.264 + 0.0876j | d3 | 0.0606 − 0.080j |
| | a4 | 0.272 + 0.0008j | d4 | 0.130 + 0.173j |
| | a6 | 0.273 − 0.0003j | d6 | 0.0106 − 0.0141j |
| | a6 | 0.27 + 0.00001j | d6 | 0.0117 + 0.0168j |
| | a7 | 0.283 − 0.009j | d7 | 0.0101 − 0.0136j |
| | a8 | 0.281 − 0.012j | d8 | 0.0063 + 0.008j |
| | a9 | 0.291 − 0.0268j | d9 | 0.0113 − 0.014j |

**Table 6.14**  Result for healthy motor i.e. under normal condition for phase B

| Condition | Approximate coefficient | R.M.S. value | Detail coefficient | R.M.S. value |
|---|---|---|---|---|
| For current of phase B under normal condition | a1 | 0.267 + 0.211j | d1 | 0.0006 − 0.0011j |
| | a2 | 0.267 + 0.210j | d2 | 0.006 + 0.010j |
| | a3 | 0.269 + 0.216j | d3 | 0.049 − 0.081j |
| | a4 | 0.261 + 0.166j | d4 | 0.093 + 0.166j |
| | a6 | 0.264 + 0.174j | d6 | 0.029 − 0.048j |
| | a6 | 0.244 + 0.132j | d6 | 0.039 + 0.064j |
| | a7 | 0.236 + 0.146j | d7 | 0.0162 − 0.0268j |
| | a8 | 0.209 + 0.101j | d8 | 0.027 + 0.044j |
| | a9 | 0.202 + 0.116j | d9 | 0.0083 − 0.013j |

**Table 6.15**  Result for 5 % winding short condition of phase B

| Condition | Approximate coefficient | R.M.S. value | Detail coefficient | R.M.S. value |
|-----------|------------------------|--------------|--------------------|--------------|
| For current of phase B under 5 % winding short | a1 | 0.2716 + 0.231j | d1 | 0.0009 − 0.0012j |
| | a2 | 0.2714 + 0.230j | d2 | 0.0082 + 0.0109j |
| | a3 | 0.286 + 0.240j | d3 | 0.0667 − 0.0886j |
| | a4 | 0.260 + 0.181j | d4 | 0.1276 + 0.1693j |
| | a6 | 0.2641 + 0.191j | d6 | 0.0401 − 0.063j |
| | a6 | 0.2343 + 0.143j | d6 | 0.06291 + 0.070j |
| | a7 | 0.223 + 0.168j | d7 | 0.022 − 0.029j |
| | a8 | 0.1876 + 0.110j | d8 | 0.0369 + 0.048j |
| | a9 | 0.1770 + 0.126j | d9 | 0.0114 − 0.016j |

**Table 6.16**  Result for 10 % winding short condition of phase B

| Condition | Approximate coefficient | R.M.S. value | Detail coefficient | R.M.S. value |
|-----------|------------------------|--------------|--------------------|--------------|
| For current of phase B under 10 % winding short | a1 | 0.260 + 0216j | d1 | 0.0007 − 0.0011j |
| | a2 | 0.260 + 0.216j | d2 | 0.0066 + 0.0102j |
| | a3 | 0.272 + 0.221j | d3 | 0.0629 − 0.083j |
| | a4 | 0.260 + 0.169j | d4 | 0.101 + 0.169j |
| | a6 | 0.264 + 0.179j | d6 | 0.0318 − 0.049j |
| | a6 | 0.241 + 0.136j | d6 | 0.0419 + 0.066j |
| | a7 | 0.233 + 0.149j | d7 | 0.0176 − 0.0276j |
| | a8 | 0.206 + 0.104j | d8 | 0.029 + 0.046j |
| | a9 | 0.196 + 0.118j | d9 | 0.009 − 0.014j |

**Table 6.17**  Result for 15 % winding short condition of phase B

| Condition | Approximate coefficient | R.M.S. value | Detail coefficient | R.M.S. value |
|-----------|------------------------|--------------|--------------------|--------------|
| For current of phase B under 15 % winding short | a1 | 0.2632 + 0.221j | d1 | 0.00079 − 0.0012j |
| | a2 | 0.2631 + 0.221j | d2 | 0.0070 + 0.0106j |
| | a3 | 0.2766 + 0.228j | d3 | 0.067 − 0.086j |
| | a4 | 0.2606 + 0.173j | d4 | 0.1090 + 0.162j |
| | a6 | 0.2644 + 0.183j | d6 | 0.0343 − 0.061j |
| | a6 | 0.2396 + 0.138j | d6 | 0.0462 + 0.0673j |
| | a7 | 0.230 + 0.162j | d7 | 0.0189 − 0.028j |
| | a8 | 0.1998 + 0.106j | d8 | 0.0316 + 0.047j |
| | a9 | 0.1907 + 0.120j | d9 | 0.00976 − 0.014j |

**Table 6.18**  Result for 20 % winding short condition of phase B

| Condition | Approximate coefficient | R.M.S. value | Detail coefficient | R.M.S. value |
|---|---|---|---|---|
| For current of phase B under 20 % winding short | a1 | 0.266 + 0.226j | d1 | 0.0008 − 0.0012j |
| | a2 | 0.266 + 0.226j | d2 | 0.0076 + 0.010j |
| | a3 | 0.280 + 0.234j | d3 | 0.0616 − 0.086j |
| | a4 | 0.260 + 0.177j | d4 | 0.117 + 166j |
| | a6 | 0.264 + 0.187j | d6 | 0.037 − 0.0621j |
| | a6 | 0.237 + 0.141j | d6 | 0.0489 + 0.0687j |
| | a7 | 0.226 + 0.166j | d7 | 0.020 − 0.028j |
| | a8 | 0.194 + 0.108j | d8 | 0.0341 + 0.048j |
| | a9 | 0.184 + 0.123j | d9 | 0.010 − 0.0141j |

**Table 6.19**  Result for healthy motor, i.e., under normal condition of phase C

| Condition | Approximate coefficient | R.M.S. value | Detail coefficient | R.M.S. value |
|---|---|---|---|---|
| For current of phase C under normal condition | a1 | 0.211 − 0.162j | d1 | 0.001 − 0.001j |
| | a2 | 0.212 − 0.162j | d2 | 0.007 + 0.011j |
| | a3 | 0.204 − 0.144j | d3 | 0.046 − 0.077j |
| | a4 | 0.266 − 0.171j | d4 | 0.093 + 0.166j |
| | a6 | 0.268 − 0.169j | d6 | 0.032 − 0.063j |
| | a6 | 0.299 − 0.134j | d6 | 0.043 + 0.071j |
| | a7 | 0.299 − 0.136j | d7 | 0.018 − 0.030j |
| | a8 | 0.326 − 0.091j | d8 | 0.029 + 0.048j |
| | a9 | 0.327 − 0.091j | d9 | 3.689e-04 − 6.19e-04j |

**Table 6.20**  Result for 5 % winding short condition of phase C

| Condition | Approximate coefficient | R.M.S. value | Detail coefficient | R.M.S. value |
|---|---|---|---|---|
| For current of phase C under 5 % winding short | a1 | 0.209 − 0.163j | d1 | 0.00116 − 0.0018j |
| | a2 | 0.210 − 0.163j | d2 | 0.00774 + 0.012j |
| | a3 | 0.200 − 0.144j | d3 | 0.0601 − 0.079j |
| | a4 | 0.266 − 0.176j | d4 | 0.101 + 0.169j |
| | a6 | 0.268 − 0.173j | d6 | 0.0347 − 0.0649j |
| | a6 | 0.301 − 0.138j | d6 | 0.0463 + 0.0732j |
| | a7 | 0.301 − 0.139j | d7 | 0.0196 − 0.0309j |
| | a8 | 0.331 − 0.093j | d8 | 0.0316 + 0.0498j |
| | a9 | 0.331 − 0.094j | d9 | 3.87e-04 − 6.34e-04j |

**Table 6.21** Result for 10 % winding short condition of phase C

| Condition | Approximate coefficient | R.M.S. value | Detail coefficient | R.M.S. value |
|---|---|---|---|---|
| For current of phase C under 10 % winding short | a1 | 0.2071 − 0.163j | d1 | 0.0013 − 0.0018j |
| | a2 | 0.2077 − 0.163j | d2 | 0.0090 + 0.0126j |
| | a3 | 0.1941 − 0.141j | d3 | 0.068 − 0.0824j |
| | a4 | 0.2668 − 0.183j | d4 | 0.117 + 0.166j |
| | a6 | 0.2686 − 0.179j | d6 | 0.0406 − 0.067j |
| | a6 | 0.3068 − 0.144j | d6 | 0.06406 + 0.076j |
| | a7 | 0.3063 − 0.146j | d7 | 0.0228 − 0.0322j |
| | a8 | 0.3410 − 0.097j | d8 | 0.0367 + 0.062j |
| | a9 | 0.3412 − 0.098j | d9 | 4.63e-04 − 6.62e-04j |

**Table 6.22** Result for 15 % winding short condition of phase C

| Condition | Approximate coefficient | R.M.S. value | Detail coefficient | R.M.S. value |
|---|---|---|---|---|
| For current of phase C under 15 % winding short | a1 | 0.2071 − 0.163j | d1 | 0.0013 − 0.0018j |
| | a2 | 0.2077 − 0.163j | d2 | 0.0090 + 0.0126j |
| | a3 | 0.1941 − 0.141j | d3 | 0.068 − 0.0824j |
| | a4 | 0.2668 − 0.183j | d4 | 0.117 + 0.166j |
| | a6 | 0.2686 − 0.179j | d6 | 0.0406 − 0.067j |
| | a6 | 0.3068 − 0.144j | d6 | 0.06406 + 0.076j |
| | a7 | 0.3063 − 0.146j | d7 | 0.0228 − 0.0322j |
| | a8 | 0.3410 − 0.097j | d8 | 0.0367 + 0.062j |
| | a9 | 0.3412 − 0.098j | d9 | 4.63e-04 − 6.62e-04j |

**Table 6.23** Result for 20 % winding short condition of phase C

| Condition | Approximate coefficient | R.M.S. value | Detail coefficient | R.M.S. value |
|---|---|---|---|---|
| For current of phase C under 20 % winding short | a1 | 0.2084 − 0.163j | d1 | 0.00124 − 0.0018j |
| | a2 | 0.2089 − 0.164j | d2 | 0.0083 + 0.012j |
| | a3 | 0.1974 − 0.143j | d3 | 0.0640 − 0.08j |
| | a4 | 0.2660 − 0.179j | d4 | 0.1091 + 0.162j |
| | a6 | 0.2683 − 0.176j | d6 | 0.0376 − 0.066j |
| | a6 | 0.3040 − 0.141j | d6 | 0.060 + 0.074j |
| | a7 | 0.3037 − 0.142j | d7 | 0.0211 − 0.0316j |
| | a8 | 0.3369 − 0.0968j | d8 | 0.0340 + 0.0609j |
| | a9 | 0.3361 − 0.096j | d9 | 4.189e-04 − 6.48e-04j |

### 6.8.3   Observations

The values of r.m.s. of approximate coefficients and detail coefficients are plotted with respect to different percentage of winding short circuit condition of the stator of the motor. These are shown in Figs. 6.19, 6.20, 6.21, 6.22, 6.23, and 6.24. From Figs. 6.19 and 6.20, it is observed that the r.m.s. values of the approximate and detail coefficients are increasing in nature for phase A. From Figs. 6.21 and 6.24 it is seen that r.m.s. values of 1st approximate coefficients to 4th approximate coefficients are increasing in nature for phase B but decreasing in nature for phase C whereas r.m.s. values of 6th approximate coefficients are constant for both phase B and C. The r.m.s. values of 6th–9th approximate coefficients are decreasing in nature for phase B but increasing in nature for phase C.

From Figs. 6.22 and 6.24 it is observed that r.m.s. values of all detail coefficients are decreasing in nature for phase B and phase C. Also it is observed that in all the figures, variation in r.m.s. values of 3rd approximate coefficients and 4th detail coefficients are very prominent. So by these two parameters, we can comment on the values of percentage of stator winding short circuit condition of an induction motor.

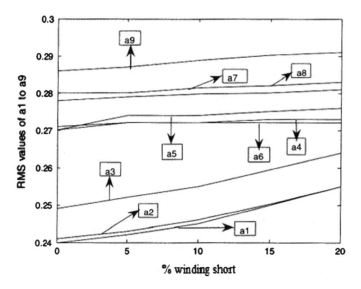

**Fig. 6.19** r.m.s. values of approximate coefficients against different percentage of winding short in phase A

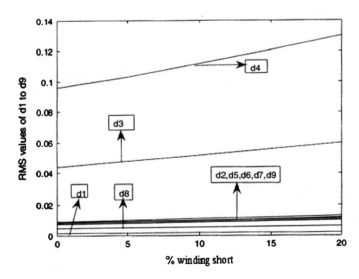

**Fig. 6.20**  r.m.s. values of detail coefficients against different percentage of stator winding short in phase A

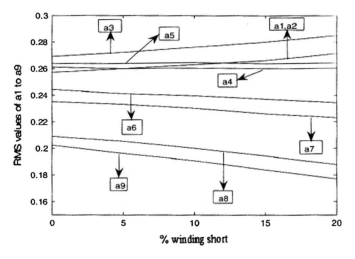

**Fig. 6.21**  r.m.s. values of approximate coefficients against different percentage of stator winding short in phase B

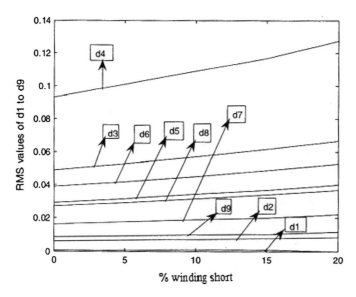

**Fig. 6.22** r.m.s. values of detail coefficients against different percentage of stator winding short in phase B

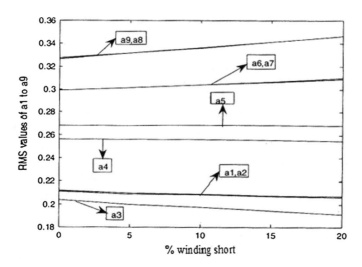

**Fig. 6.23** r.m.s. values of approximate coefficients against different percentage of stator winding short in phase C

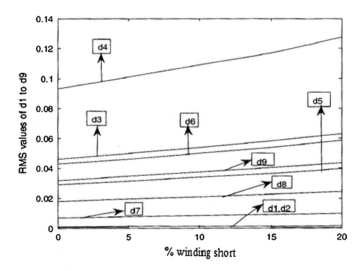

**Fig. 6.24**  r.m.s. values of detail coefficients against different percentage of stator winding short in phase C

## 6.9  Conclusion

In this chapter, stator winding fault of an induction motor has been diagnosed using different techniques. First, the conventional methods and then modern signal processing tool like Wavelet Transform has been used to diagnose the fault.

First negative sequence component of stator current is used. Change of negative sequence component of current for different percentage short of stator winding is observed. An algorithm is developed for the detection purpose. In the second technique, FFT is performed on the steady portion of the stator current. It is shown that due to change in the percentage short of stator winding, amplitude of sidebands of the fundamental harmonic changes very prominently; total harmonic distortion (THD) also changes accordingly.

In the third technique, analysis of both steady and transient parts of the stator currents is performed. For the steady part of the current, wave energy is calculated which shows distinct variation in 6th, 6th and 7th decomposition level for different percentage of stator winding short. In the fourth technique, Continuous Wavelet Transform (CWT) is performed on the transient part of stator current and statistical parameters namely; Kurtosis and Skewness values of the CWT coefficients are used to diagnose the fault. At last, DWT is performed and r.m.s. values of both detail and approximate coefficients are used to diagnose the fault.

# References

1. Singh GK, Al Kazzaz SAS (2003) Induction machine drive condition monitoring and diagnostic research—a survey. Electr Power Syst Res 64(2):146–168
2. Allbrecht PF, Appiarius JC, McCoy RM, Owen EL (1986) Assessment of the reliability of motors in utility applications—updated. IEEE Trans Energy Convers EC-1(1):39–46
3. Lee SB, Tallam RM, Habetler TG (2003) A robust on-line turn-fault detection technique for induction machines based on monitoring the sequence component impedance matrix. IEEE Trans Power Electron 18(3):866–872
4. Siddique A, Yadava GS, Singh B (2006) A review of stator fault monitoring techniques of induction motors. IEEE Trans Energy Convers 20(1):106–114
5. Karmakar S, Chattopadhyay S, Mitra M, Sengupta S (2013) Inter turn short circuit assessment of an induction motor using negative sequence component. In: Proceedings of national conference on recent developments in electrical, electronics & engineering physics (RDE3P)—2013, MCKVIE, pp 36–37, October 26–27, 2013. ISBN: 978-81-8424-877-7
6. Chattopadhyay S, Mitra M, Sengupta S (2011) Electric power quality, 1st edn. Springer, New York
7. Karmakar S, Chattopadhyay S, Mitra M, Sengupta S (2013) Inter-turn short circuit fault diagnosis of an induction motor by FFT of stator current, Michael Faraday IET India Summit (MFIIS-2013), pp 6.26–6.29, 17 Nov 2013. ISBN: 978-93-82716-97-9
8. Karmakar S, Chattopadhyay S, Mitra M, Sengupta S (2014) Turn-to-turn fault diagnosis of an induction motor by the analysis of transient and steady state stator current. J Innovative Syst Des Eng. ISSN 2222-1727 (paper) ISSN 2222-2871(Online), IISTE, vol 6, No 2, pp 66–74
9. Gaouda AM, Salma AMM, Sultan MR, Chikhani AY (1999) Power quality detection and classification using wavelet-multiresolution signal decomposition. IEEE Trans Power Delivery 14(4):1469–1476
10. Karmakar S, Chattopadhyay S, Mitra M, Sengupta S (2012) Stator winding fault assessment of an induction motor by starting current transient analysis, Michael Faraday IET India Summit by IET Kolkata Network, Kolkata, pp 266–268
11. Karmakar S, Ghosh S, Banerjee H, Chattopadhyay S (2013) Wavelet transformation based stator current analysis for short circuit analysis of an induction motor. In: Proceedings of national conference on recent developments in electrical, electronics & engineering physics (RDE3P)—2013, MCKVIE, pp 69–64, 26–27 Oct 2013. ISBN: 978-81-8424-877-7
12. Lipo TA (2004) Introduction of AC machine design, 2nd edn. Wisconsin Power Electronics Research Center, Madison

# Chapter 7
# Single Phasing of an Induction Motor

**Abstract** This chapter deals with the diagnosis of single phasing fault in induction motor. Motor current signature analysis (MCSA) technique has been discussed. Phase angle shift is assessed at single phasing. Steady-state current Concordia is formed and assessed by feature pattern extraction method (FPEM). Then, radar analysis of line currents and those Concordia during normal and single phasing are discussed.

**Keywords** CMS rule set · Concordia · FPEM · Phase angle shift · Radar · Single phasing

**Chapter Outcome**

After completion of the chapter, readers will be able to gather knowledge and information regarding the following areas:

- Single phasing
- Signature analysis
- Phase shifting
- Concordia
- Radar-based diagnosis
- Diagnosis using feature pattern extraction method.

## 7.1 Introduction

Single phasing fault is a power supply-related electrical fault. When any of the three phases of an induction motor gets disconnected then the phenomenon is known as single phasing fault. If single phasing fault occurs, then the current components of the remaining two phases change along with the flow of negative sequence components of line currents. Hence, by analyzing the stator current signature, single phasing fault can be diagnosed. In this chapter, first a brief description of the single

© Springer Science+Business Media Singapore 2016
S. Karmakar et al., *Induction Motor Fault Diagnosis*,
Power Systems, DOI 10.1007/978-981-10-0624-1_7

phasing fault will be given. Negative sequence component and detection of its magnitude will be discussed and then the methods of diagnosis of this single phasing fault will be described. The chapter will be concluded by a comparison of the diagnosing methods and their advantages and disadvantages.

## 7.2  Single Phasing Fault

In induction motor, faults that occur can be broadly divided into two groups—electrical fault and mechanical fault. Single phasing fault is one kind of electrical fault related to power supply. For a three-phase motor, when one of the three phases is damaged then the condition created is called single phasing. Due to single phasing fault various types of effects may occur in an induction motor. If a single phasing fault exists during starting the motor then it will not get started. During running condition of the motor if single phasing occurs, the motor continues to run and if it runs with this fault for a long time or with a heavy load then the remaining two phases may also get damaged. Due to this fault, induction motor windings get overheated.

## 7.3  Diagnosis of Single Phasing Fault

For diagnosis of single phasing fault in induction motor, motor current signature analysis (MCSA) technique has been used in the present work. Assessment of phase angle shift, current Concordia analysis, radar analysis of line currents, and stator current Concordia by feature pattern extraction method (FPEM) are performed to diagnose the single phasing fault in induction motor as shown in Fig. 7.1. These methods will be discussed in subsequent sections.!

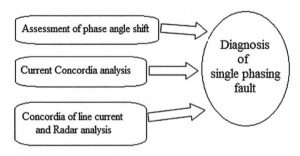

**Fig. 7.1**  Diagnosis of single phasing fault

## 7.4 Single Phasing Fault Detection by Assessment of Phase Angle Shift [1, 2]

### 7.4.1 Theory

The angular relations among sequence components when single phasing occurs due to loss of R phase are as follows [3]:

$$I_{R1} = I_{R2} \angle 182°$$

$$I_{Y2} = I_{Y1} \angle 60°$$

$$I_{B1} = I_{B2} \angle 60°$$

The angular relations among sequence components when single phasing occurs due to loss of Y phase are as follows [3]

$$I_{R1} = I_{R2} \angle 60°$$

$$I_{Y1} = I_{Y2} \angle 180°$$

$$I_{B2} = I_{B1} \angle 60°$$

The angular relations among sequence components when single phasing occurs due to loss of B phase are as follows [3]

$$I_{Y1} = I_{Y2} \angle 60°$$

$$I_{Y1} = I_{Y2} \angle 60°$$

$$I_{B1} = I_{B2} \angle 180°$$

From the above, it is observed that the negative sequence components are at an angle of 60° to that of the positive sequence components in healthy lines and 180° out of phase to that of positive sequence components at the line where loss of phase has occurred.

### 7.4.2 Experimentation

For experimentation, a three-phase squirrel-cage induction motor, as given in Table 7.1, is considered. The motor is connected with three-phase power supply through three variable reactors as shown in Fig. 7.2, and is operated at no load. Line currents drawn by the motor are stepped down by three current transformers (CT). Secondary sides of these CT are connected with resistance. Resistances are adjusted

**Table 7.1**  Specification of the motor used for experiment

| Make | M/S Ventwell Corporation |
|------|--------------------------|
| Power | 5HP (3.7 kW) |
| Voltage | 415 V |
| Frequency | 50 Hz |
| Speed | 1425 rpm |

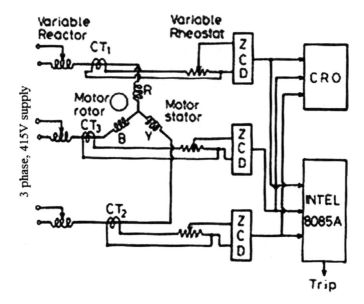

**Fig. 7.2**  Block diagram of the experimental set-up

in such a way that the effect of CT saturation does not appear during the experiment. Voltage across each resistance, which is the measure of line current, is sent to zero crossing detectors (ZCD). ZCD converts sinusoidal waves into square waves which were taken as inputs for an 8085A Intel microprocessor. The magnitudes of the reactors are initially kept very small, then, one (say, reactor of phase B) is gradually increased. As a result, the current through phase B gradually reduces to zero causing loss of phase in phase B.

### 7.4.3  Results and Discussions

The phase angle difference between two healthy phases (R and Y) shifts from 120° to 180° and is observed in the CRO. This angle shift is measured by the microprocessor by measuring the time difference between rising edges of the two healthy waveforms of phase R and phase Y.

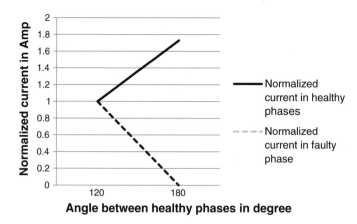

**Fig. 7.3** Normalized current versus angle between currents through healthy phases

From experimentation, angular shift of phase current during single phasing and negative sequence component of current are calculated and shown in Fig. 7.3. It shows almost linear relationship.

## 7.5   Detection of Single Phasing by Steady-State Current Concordia and Radar Analysis [1]

### 7.5.1   Steady-State Concordia Analysis Using CMS Rule Set

Let $x(t)$ and $y(t)$ be two sinusoids separated by $\phi$.

$$x(t) = X \sin \omega t \tag{7.1}$$

$$y(t) = Y \sin(\omega t - \phi) \tag{7.2}$$

In $x$-$y$ plane these two waves form elliptical Concordia as shown in Fig. 7.4. Major and semi-major axes of this Concordia depend on the amplitudes of $x(t)$ and $y(t)$.

With this theoretical background, assessment of currents drawn by an induction motor has been done here by the analysis of Concordia formed by motor currents. Now, stator currents drawn by an induction motor can be written as

$$i_R = I_R \sin \omega t \tag{7.3}$$

$$i_Y = I_Y \sin(\omega t - 120°) \tag{7.4}$$

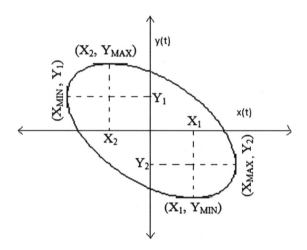

**Fig. 7.4**  Concordia formed by $x(t)$ versus $y(t)$ in $x$-$y$ plane

$$i_B = I_B \sin(\omega t - 240°) \tag{7.5}$$

Harmonics of the system (if any) are to be filtered out. For balanced load, amplitudes are equal, i.e.,

$$I_R = I_Y = I_B$$

Eliminating $\omega t$ term currents can be represented as

$$i_R = I_R \sin\left\{ \sin^{-1}\left(\frac{i_B}{I_B}\right) + 240° \right\} \tag{7.6}$$

$$i_Y = I_R \sin\left\{ sin^{-1}\left(\frac{i_R}{I_R}\right) - 120° \right\} \tag{7.7}$$

$$i_B = I_R \sin\left\{ \sin^{-1}\left(\frac{i_Y}{I_Y}\right) - 120° \right\} \tag{7.8}$$

Equations (7.6), (7.7), and (7.8) generate elliptical Concordia in current-current planes.

For current assessment, following parameters shown in Fig. 7.4, are introduced as follows:

- at $x(t) = X_{MIN}$, $y(t) = Y_1$
- at $x(t) = X_{MAX}$, $y(t) = Y_2$
- $y = Y_1 \sim Y_2$
- at $y(t) = Y_{MIN}$, $x(t) = X_1$
- at $y(t) = Y_{MAX}$, $x(t) = X_2$

- $x = X_1 \sim X_2$

$x$ and $y$ are the measure of two amplitudes.

Three Concordia are formed corresponding to phase-combinations R-Y, Y-B, and B-R. For each combination '$x$' and '$y$' are calculated and thus two matrices are formed as follows:

$$[x] = \begin{bmatrix} x_{RY} \\ x_{YB} \\ x_{BR} \end{bmatrix} \quad \text{and} \quad [y] = \begin{bmatrix} y_{RY} \\ y_{YB} \\ y_{BR} \end{bmatrix} \tag{7.9}$$

The values of $[x]$ and $[y]$ at different conditions like balanced, unbalanced, and fault situations have been calculated. Using these matrices $[x]$ and $[y]$, CMS rule set [3] are used for unbalance monitoring.

Two important rules of CMS rule set are

(a)  If $[x] = [1\ 1\ 1]$ and $[y] = [1\ 1\ 1]$, then the power system is perfectly balanced.
(b)  If $[x] \neq [1\ 1\ 1]$ and $[y] \neq [1\ 1\ 1]$, then power system is unbalanced.

In case of unbalanced situation, a generalized inference are drawn as follows:

- $x_{RY}$ and $y_{BR}$ are the measure of current through phase R
- $x_{YB}$ and $y_{RY}$ are the measure of current through phase Y
- $x_{BR}$ and $y_{YB}$ are the measure of current through phase B

Concordia made of one healthy phase and the faulty phase lies either on horizontal axis or on vertical axis whereas Concordia formed by two healthy phase inclines in second and third quadrant of the plane. Based on the above inference, logic is developed for loss of phase fault detection as follows:

If $[x] = \begin{bmatrix} \text{Zero} \\ \text{High} \\ \text{High} \end{bmatrix}$ and $[y] = \begin{bmatrix} \text{High} \\ \text{High} \\ \text{Zero} \end{bmatrix}$ then, loss of phase has occurred in phase R.

If $[x] = \begin{bmatrix} \text{High} \\ \text{Zero} \\ \text{High} \end{bmatrix}$ and $[y] = \begin{bmatrix} \text{Zero} \\ \text{High} \\ \text{High} \end{bmatrix}$ then, loss of phase has occurred in phase Y.

If $[x] = \begin{bmatrix} \text{High} \\ \text{High} \\ \text{Zero} \end{bmatrix}$ and $[y] = \begin{bmatrix} \text{High} \\ \text{Zero} \\ \text{High} \end{bmatrix}$ then, loss of phase has occurred in phase B.

## 7.5.2   Radar Analysis [1, 2]

Stator currents of the induction motor are stepped down through three CT and captured by data acquisition system (DAS). Then, for experimentation, loss of phase is implemented on each of three phases R, Y and B, separately. Concordia are formed corresponding to healthy condition of the motor and loss of phase at different phases, one after another. Then matrices $[x]$ and $[y]$ (discussed in previous

**Table 7.2** Concordia and line current radar diagram of [*x*] at healthy condition

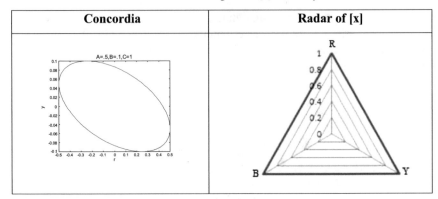

| Concordia | Radar of [x] |
|---|---|

section) are calculated from these Concordia and radars are formed for [*x*]. Stator current Concordia and corresponding Radars during normal condition are presented in Table 7.2. It shows that when the motor is in healthy condition, Concordia is elliptical in shape and radars of [*x*] are of triangular in shape.

Then, stator current Concordia are formed and [*x*] is calculated during single phasing. After this Radar of [*x*] is formed. Stator current Concordia and corresponding Radars during single phasing at phases R, Y and B respectively are presented in Table 7.3. It shows that all Concordia are straight line during single phasing. Concordia made of one healthy phase and the faulty phase lies either on the horizontal axis or on the vertical axis whereas Concordia formed by two healthy phases inclines in the second and third quadrants of the plane. Radars of [*x*] are of triangular shape and they are located at the planes of two healthy phases.

*Inference*:

Inference can be drawn for diagnosis of single phasing as follows:

- During single phasing, magnitude of line currents of healthy phases increase by $\sqrt{3}$ times and phase angle difference between currents of two healthy phases becomes 180°.
- During single phasing elliptical Concordia which are made of two consecutive stator current convert into straight Concordia; Concordia made of healthy line currents make 45° with horizontal and vertical axis.
- At normal condition, radar of [*x*] forms triangle made of three equal arms.
- During single phasing, radar of [*x*] forms triangle which has two of equal arms and the triangle lies on planes of healthy phases.

In this work, single phasing is first assessed by conventional sequence component methods where negative sequence components and angle shift of healthy phases are measured. Then stator current Concordia is formed and the rule set is presented for detection of single phasing. Concordia of stator current during single phasing is straight line. Matrix [*x*] is measured and the radars of [*x*] are formed.

**Table 7.3** Stator current Concordia and radar diagram of [x] for single phasing at phase R, Y, and B

| Single phasing at phase | Concordia | Radar of [x] |
|---|---|---|
| R | | |
| Y | | |
| B | | |

These radars are of triangular shape. The radar of [x] during single phasing lies on the plane of healthy phases.


## 7.6   Conclusion

In this chapter, single phasing fault in an induction motor has been diagnosed in different ways.

Conventional sequence component method is discussed, where negative sequence components and angle shift of stator current are measured. A graph of angular shift of phase current during single phasing versus the negative sequence component of stator current is shown in Fig. 7.3.

Current Concordia analysis by CMS rule set—when all phases are present pattern formed by any two phases becomes elliptical. But when there is loss of phase, the corresponding pattern becomes straight line. Then CMS set is applied for diagnosis of single phasing.

Concordia at single phasing—for healthy motor, Concordia is elliptical in shape and radars of [x] are of equilateral triangle in shape. Whereas incase of motor with single phasing, Concordia is a straight line and the radars of [x] are of triangular in shape but not equilateral and the triangles are located at the planes of two healthy phases indicating the phase location of the fault.


## References

1. Chattopadhyay S, Karmakar S, Mitra M, Sengupta S (2012) Loss of phase fault detection of an induction motor. Int J Model Meas Control Gen Phys Electr Appl AMSE. Series A, 85(2):18–34. ISSN 1259-5985
2. Chattopadhyay S, Karmakar S, Mitra M, Sengupta S (2012) Symmetrical components and current Concordia based assessment of single phasing of an induction motor by feature pattern extraction method and Radar analysis. Int J Electr Power Eng Syst 37(1):43–49. ISSN 0142–0615
3. Chattopadhyay S, Mitra M, Sengupta S (2011) Electric power quality, 1st edn. Springer, Berlin

# Chapter 8
# Crawling of an Induction Motor

**Abstract** This chapter deals with the assessment of crawling. The assessment is done in Park plane. Steady-state stator currents are transformed into Park plane. Concordia is formed in the Park plane. Presence of 7th harmonics is assessed by the feature pattern extraction method (FPEM) and the CMS rule set used for harmonic assessment in Park plane from where crawling is detected.

**Keywords** CMS rule set · Concordia · Crawling · FPEM · Park plane · Seventh harmonics

**Chapter Outcome**

After completion of the chapter, readers will be able to gather knowledge and information regarding the following areas:

- Crawling
- Harmonic Assessment in Park plane
- Diagnosis using CMS rule set.

## 8.1 Introduction

If an induction motor under loaded condition does not accelerate up to its normal speed but runs at a speed nearly one-seventh of the synchronous speed, then it is said that the induction motor is crawling. In this chapter, crawling—its causes and how it can be detected is studied. At the end, a conclusion is made on this chapter.

## 8.2 Crawling in Induction Motor

In a three-phase induction motor the rotating air gap flux in between stator and rotor contains odd harmonics. Out of these, flux due to third harmonics and its multiples produced by each of the three phases neutralize each other also magnitudes of 11th and

© Springer Science+Business Media Singapore 2016

S. Karmakar et al., *Induction Motor Fault Diagnosis*, Power Systems, DOI 10.1007/978-981-10-0624-1_8

higher order harmonics being very small only 5th and 7th harmonics are predominant. All harmonic fluxes rotate at $N_s/k$ rpm speed ($k$ denotes the order of the harmonics, for fundamental $k = 1$), in the same direction of the fundamental except the fifth harmonic. Flux due to fifth harmonic rotates in opposite direction to the fundamental flux.

Now from the construction and operation (Chap. 2) of an induction motor, it is known that motor torque is produced due to the air gap flux and total motor torque has three components—(i) fundamental torque, rotating at synchronous speed $N_s$, (ii) 5th harmonic torque, rotating at speed $N_s/5$ in the opposite direction of fundamental, and (iii) 7th harmonic torque, rotating at speed $N_s/7$ in the same direction of fundamental. The 5th harmonic torque produces a breaking action of small magnitude and hence can be neglected. Thus, the resultant torque is considered as the sum of the fundamental torque and the 7th harmonic torque [1, 2]. This sum shows a dip at speed of $n_s/7$ which results in decrease in torque even when the speed increases due to which the induction motor will remain running at a speed of nearly at $n_s/7$. This situation is termed as crawling.

Crawling is predominant in the squirrel-cage type induction motor. It causes higher stator current, lowering of speed, vibration, and noise in the motor. By proper design of the stator winding the crawling effect can be reduced. Harmonics being the main reason of crawling in an induction motor, it can be diagnosed through the identification of the harmonics present in the stator current.

## 8.3  Diagnosis of Crawling of an Induction Motor by Feature Pattern Extraction of Stator Current Concordia [3]

### 8.3.1  Concordia in Park Plane

Currents in direct–quadrature ($d$-$q$) plane can be obtained from phase current multiplied by Park matrix as follows:

$$\begin{pmatrix} i_d \\ i_q \end{pmatrix} = (\text{Park Matrix}) \times \begin{pmatrix} i_a \\ i_b \\ i_c \end{pmatrix} \tag{8.1}$$

where,

$$(\text{Park Matrix}) = \sqrt{\frac{2}{3}} \times \begin{pmatrix} \cos\theta & \cos\left(\theta - \frac{2\pi}{3}\right) & \cos\left(\theta - \frac{4\pi}{3}\right) \\ -\sin\theta & -\sin\left(\theta - \frac{2\pi}{3}\right) & -\sin\left(\theta - \frac{4\pi}{3}\right) \end{pmatrix}$$

Concordia formed by these currents looks circular in Park plane. In this plane, harmonics-free stator current form exact circular Concordia. But the presence of harmonics brings cleavages in this Concordia.

## 8.3.2   Pattern Generation and Inference

Concordia is generated with signals having of different order of harmonics in Park plane. Figure 8.1 shows three patterns formed by *d* axis and *q* axis currents. In this figure, the pattern free from cleavage corresponds to a balanced harmonic free system, the pattern with four cleavages corresponds to system with fifth-order harmonic and the other pattern with five cleavages corresponds to the system with sixth-order harmonic.

## 8.3.3   CMS Rule Set for Harmonic Assessment in Park Plane

From the analysis of above Concordia, CMS rule set is used for identifying the presence of the single harmonic present in the system.

Rule 1: If cleavage appears at the Concordia in Park plane, then there is harmonic in the system

Rule 2: If rule 1 is true and if the number of cleavages is *C* and order of harmonic is *n*, then $n = C + 1$

Rule 3: If rule 1 is true, then there will be at least one cleavage at an angle $\theta$ given by $\alpha_n = 270° + \theta/2$ for odd order (*n*) and $\alpha_n = 270° + \theta$ for even order (*n*); where, $\theta = 360°/C$

Rule 4: If rules 1, 2, and 3 are true, then the percentage of harmonic is proportional to depth of locus.

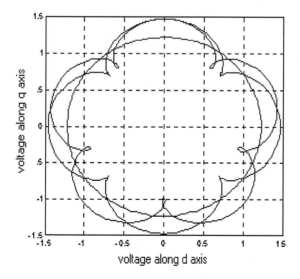

**Fig. 8.1**   Concordia generated by fundamental, fundamental with fifth-order and fundamental with sixth-order harmonics in Park plane

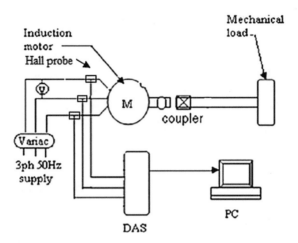

**Fig. 8.2**  Block diagram of the experimental set-up

*Example 8.1* The block diagram of the experimental setup is shown in Fig. 8.2. Three phase 110 V, 50 Hz, AC supply is provided to the induction motors (1/3 HP, wound rotor) that suffers from crawling. The motor is run by direct-on-line supply. Hall Probe (LEM PR30 ACV 600 V CATIII 30 Ampac/3 Vac) is used for data collection. Stator current signal has been captured with a sample frequency of 5120 Hz. The assessment will be discussed in the following section by the data obtained from this example.

Details about fault detection are discussed in this section with the help of results and observation obtained by experimentation with this Example 8.1.

### 8.3.4   *Results and Discussions*

Three-phase stator currents are normalized and transformed into Park plane. Then, Concordia is formed in the Park plane as shown in Fig. 8.3 from where the nature of cleavages is observed. Six numbers of cleavages are observed which indicate that seventh-order harmonics are dominating. Depth of the cleavage is almost proportional to the percentage of the harmonics with respect to fundamental. The result shows that the number and depth of cleavages of the Concordia formed in the Park plane is capable of giving information about the harmonics present in the stator current.

Thus, crawling has been analyzed by assessing generated harmonics using stator current Concordia. Stator current has been captured, normalized, and then Concordia has been formed in Park plane. Harmonics have been assessed by rule set based on the features extracted from the patterns developed by stator currents. Dominance of 7th harmonic which is the main cause of crawling is determined without using any filter or any Fourier- or Wavelet transform-based technique.

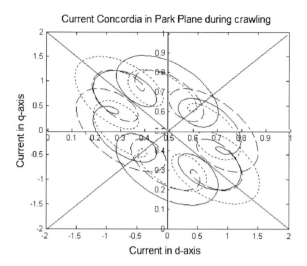

**Fig. 8.3**  Current Concordia in the Park plane during crawling

## 8.4   Conclusion

In this chapter, crawling of an induction motor has been detected by the analysis of stator current signature. A very simple method, called feature pattern extraction method (FPEM), is described where the first Concordia is formed by the stator current in the Park's plane, then by simply observing the number of cleavages present in the Concordia, dominance of the seventh harmonic is detected using CMS rule set. Now in an induction motor, the seventh harmonic being the main cause of crawling it can be detected very easily from this Concordia.

## References

1. Say MG (2002) The performance and design of alternating current machines. M/S Pitman, London. ISBN: 81-239-1027-4
2. Chattopadhyay S, Mitra M, Sengupta S (2011) Electric power quality, 1st edn. Springer, Netherlands
3. Chattopadhyay S, Karmakar S, Mitra M, Sengupta S (2012) Assessment of crawling of an induction motor by stator current Concordia analysis. IET Electron Lett 48(14):841–842. doi:10.1049/el.2011.4008

# Chapter 9
# Induction Motor Fault Diagnosis: General Discussion and Research Scope

**Abstract** This chapter deals with general discussion and research scope on induction motor fault diagnosis. Fault considered in the book **and** analytical tools used for motor fault diagnosis has been listed. Then, general discussion on induction motor fault diagnosis has been made. At last research scope in this field has been highlighted.

**Keywords** Analytical tools · CMS rule set · Concordia analysis · Continuous wavelet transform (CWT) · Discrete wavelet transform (DWT) · Fast fourier transform (FFT) · General discussion · Hilbert transform (HT) · Induction motor fault · Kurtosis · Radar · Research scope · Sequence component analysis · Total harmonic distortion · Signals for fault diagnosis · Skewness

**Chapter Outcome**

After completion of the chapter, readers will be able to gather knowledge and information regarding following areas:

- Faults considered in the book
- Analytical tools used for motor fault diagnosis
- General discussion on induction motor fault diagnosis
- Research scope.

## 9.1 Introduction

An effective and relevant literature survey has been presented in Chap. 1 on induction motor's construction and operation. Faults in induction motor are studied along with their causes and effects. In previous chapters, faults like rotor broken bar fault, air gap eccentricity fault, rotor mass unbalance fault, stator fault, single phasing fault, and crawling have been discussed. Different mathematical tools and practical procedures to deal with faults are studied. Literature survey reveals that

© Springer Science+Business Media Singapore 2016
S. Karmakar et al., *Induction Motor Fault Diagnosis*,
Power Systems, DOI 10.1007/978-981-10-0624-1_9

existing techniques used for fault analysis of induction motor are mainly thermal analysis, chemical analysis, acoustic analysis, torque analysis, induced voltage analysis, partial discharge analysis, vibration analysis, and current analysis. From these, motor current analysis is very popular. In following sections, analytical tools, signals used, general discussion on induction motor fault diagnosis and research scope have been presented.

## 9.2   Faults, Analytical Tools and Signals Used in the Book

Faults discussed in depth in this book are broken rotor bar fault, rotor mass unbalance fault, stator winding fault, single phasing, and crawling. For diagnosis of these fault both steady-state motor current and starting current are used. Steady state current is easy to capture as it exists for long period, whereas starting current is less depended on load condition. For fault diagnosis, following analytical tools are used: fast Fourier transform (FFT), continuous wavelet transform (CWT), discrete wavelet transform (DWT), Hilbert Transform (HT), sequence component analysis, CMS rule set, Concordia analysis, radar analysis, total harmonic distortion assessment, skewness and kurtosis assessment. Faults, analytical tools and signals used in the book are presented in Table 9.1.

**Table 9.1**  Faults, analytical tools, and signals used in fault diagnosis

| Faults assessed in the book | Broken rotor bar fault |
|---|---|
| | Rotor mass unbalance fault |
| | Stator winding fault |
| | Single phasing |
| | Crawling |
| Signals used in fault diagnosis | Starting current transients |
| | Steady-state current |
| Analytical tools in the book | Fast Fourier transform (FFT) |
| | Continuous wavelet transform (CWT) |
| | Discrete wavelet transform (DWT) |
| | Hilbert transform (HT) |
| | Sequence component analysis |
| | CMS rule set |
| | Concordia analysis |
| | Radar analysis |
| | Total harmonic distortion assessment |
| | Skewness and kurtosis assessment |

## 9.3 General Discussion on Induction Motor Fault Diagnosis

This book deals motor current signature analysis for fault diagnosis of three phase induction motor. Stress that has been given is on assessment of faults namely, rotor broken bar fault, rotor mass unbalance fault, stator winding fault, single phasing fault, and crawling which may occur in a three phase induction motor.

Broken rotor bar fault [1–7] can be identified by using both steady-state signal as well as starting current transients. Analytical tools that may be used for diagnosis of broken rotor bar are as follows: (i) diagnosis of the fault through radar analysis of stator current Concordia and (ii) diagnosis through envelope analysis of motor startup current using Hilbert and wavelet transform.

Rotor mass unbalance fault [4, 8–10] may be assessed by both transient and steady-state stator current using different techniques. Diagnosis can be done by FFT analysis of steady-state motor vibration and current signatures, by stator current Concordia at starting assessment, by radar analysis, by analysis of statistical parameters of DWT coefficients of starting current at no load and by the analysis of PDD of reconstructed starting current after wavelet transform. If two faults namely, rotor broken bar fault and rotor mass unbalance fault occur simultaneously then for identification radar-based algorithm may be used (described in Chap. 5).

Assessment of stator winding fault [11–15] may be done by various ways like (i) diagnosis using negative sequence component of stator current at steady state, (ii) diagnosis by FFT analysis of steady-state stator current, (iii) diagnosis by CWT analysis of transient current and FFT analysis of steady-state stator current, (iv) diagnosis by starting current transient analysis using statistical parameters like skewness, kurtosis, etc., and (v) diagnosis by wavelet transformation of stator current in Park plane.

Single phasing fault [16, 17] may be detected and analyzed by (i) assessment of phase angle shift, (ii) current Concordia analysis; (iii) CMS rule set, and (iv) radar analysis.

Crawling of an induction motor [18] may be detected by feature pattern extraction method (FPEM). Converting the phase current into $d$-$q$ plane by Park's matrix, Concordia is generated in Park plane and by analysis of the pattern, harmonics present in the stator current is determined by CMS rule set and crawling is detected.

Thus signals to be chosen for motor fault diagnosis depend on the availability and loading effect. Analytical tool is to be chosen depending on the fault type and signals used for fault diagnosis.

## 9.4    Main Achievements

Main achievements of the work may be written as follows:

- Rotor broken bar fault, mass unbalance, stator fault, single phasing fault, air gap eccentricity fault and crawling have been assessed by different methods and a comparative study has been made.
- Assessment has been done using stator current.
- Assessment is done using transient as well as steady-state stator current.
- Stator current both under no-load and load are used in assessment.
- FPEM is used for fault assessment using steady-state current.
- Sequence components have also used during unbalanced situation.
- CMS rule set has been effective used for fault diagnosis.
- Concordia analysis has been effective used in motor fault diagnosis.
- Radar analysis has been found effective and easy tool for fault diagnosis.
- FFT of the steady-state current has been used for fault diagnosis.
- Wavelet transformation of transient current has been used for fault diagnosis.
- Statistical parameters like r.m.s. value, mean value, skewness and kurtosis have been utilized for fault diagnosis.

## 9.5    Research Scope

Research is a continuous process. Though a lot of researches have been done in motor fault analysis, still wide area is available for researcher to study in-depth. Some areas may be listed as follows:

- Machine's activity-based model can be reproduced during fault conditions.
- Harmonics powers can be assessed during all fault condition.
- The work may be extended for the detection of faults of motors connected in common bus.
- Rotor broken bar fault, mass unbalance, stator fault, single phasing fault, air gap eccentricity fault, and crawling have been assessed by different methods and a comparative study has been made.
- Assessment has been done using rotor current for wound rotor motor.
- Assessment can be done using switching off transient current.
- FPEM may be used for other type of faults.
- Sequence components may be assessed by different mathematical tools.
- CMS rule set has been effectively used for diagnosis of other faults.
- Concordia analysis has been effectively used in diagnosis of other motor fault.

- Radar analysis may be done for other faults and on rotor current for wound rotor motor.
- Radar analysis may be done in Park and Clarke plane.
- Statistical parameters like r.m.s. value, mean value, skewness and kurtosis may be utilized for diagnosis of other faults.
- Activity-based model may be used for motor fault diagnosis to study the generation of harmonic due to faults.

# References

1. Deleroi W (1984) Broken bars in squirrel cage rotor of an induction motor-part I: description by superimposed fault currents. Arch Elektrotech 67:91–99
2. Filippetti F, Franceschini G, Tassoni C, Vas P (1998) AI techniques in induction machines diagnosis including the speed ripple effect. IEEE Trans Ind Appl 34:98–108
3. Chattopadhyay S, Karmakar S, Mitra M, Sengupta S (2012) Radar analysis of stator current Concordia for diagnosis of unbalance in mass and cracks in rotor bar of an squirrel cage induction motor. Int J Model, Meas Control Gen Phys Electr Appl, AMSE, Ser A 85(1):50–61. ISSN 1259-5985
4. Chattopadhyay S, Mitra M, Sengupta S (2011) Electric power quality, 1st edn. Springer, Berlin
5. Boashash B (1992) Estimating and interpreting the instantaneous frequency of a signal—part 1: fundamentals. Proc IEEE 80(4):520–538
6. Chattopadhyay S, Mitra M, Sengupta S (2007) Harmonic analysis in a three-phase system using park transformation technique. Modeling-A. AMSE Int J Model Simul 80(3):42–58
7. Ahamed SK, Karmakar S, Mitra M, Sengupta S (2011) Diagnosis of broken rotor bar fault of induction motor through envelope analysis of motor startup current using Hilbert and wavelet transform. J Innovative Syst Design Eng 2(4):163–176
8. Karmakar S, Ahamed SK, Mitra M, Sengupta S (2007) Diagnosis of fault due to unbalanced rotor of an induction motor by analysis of vibration and motor current signatures. In: International conference MS'07, India, 3–5 Dec, pp 399–403
9. Ahamed SK, Karmakar S, Mitra M, Sengupta S (2009) Detection of mass unbalance rotor of an induction motor using wavelet transform of the motor starting current at no load. In: Proceedings of national conference on modern trends in electrical engineering (NCMTEE-2009), organized by IET and HETC, Hooghly, West Bengal, 11–12 July 2009, pp MC-1–MC-6
10. Ahamed SK, Karmakar S, Mitra M, Sengupta S (2010) Novel diagnosis technique of mass unbalance in rotor of Induction motor by the analysis of motor starting current at no load through wavelet transform. In: 6th international conference on electrical and computer engineering, ICECE 2010, 18–20 Dec 2010, Dhaka, Bangladesh, pp 474–477. 978-1-4244-6279-7/10©2010. IEEE IEEE Xplore
11. Karmakar S, Chattopadhyay S, Mitra M, Sengupta S (2013) Inter turn short circuit assessment of an induction motor using negative sequence component. In: Proceedings of national conference on recent developments in electrical, electronics and engineering physics (RDE3P) —2013, MCKVIE, 26–27 Oct 2013, pp 36–37. ISBN 978-81-8424-877-7
12. Karmakar S, Chattopadhyay S, Mitra M, Sengupta S (2013) Inter-turn short circuit fault diagnosis of an induction motor by FFT of stator current. Michael Faraday IET India Summit (MFIIS-2013), 17 Nov 2013, pp 6.26–6.29. ISBN 978-93-82716-97-9

13. Karmakar S, Chattopadhyay S, Mitra M, Sengupta S (2014) Turn-to-turn fault diagnosis of an induction motor by the analysis of transient and steady state stator current. J Innov Syst Des Eng IISTE 6(2):66–74. ISSN 2222-1727 (paper) ISSN 2222-2871(Online)

14. Karmakar S, Chattopadhyay S, Mitra M, Sengupta S (2012) Stator winding fault assessment of an induction motor by starting current transient analysis. Michael Faraday IET India Summit by IET Kolkata Network, Kolkata, Nov 2012, pp 266–268

15. Karmakar S, Ghosh S, Banerjee H, Chattopadhyay S (2013) Wavelet transformation based stator current analysis for short circuit analysis of an induction motor. In: Proceedings of national conference on recent developments in electrical, electronics and engineering physics (RDE3P)—2013, MCKVIE, 26–27 Oct 2013, pp 69–64. ISBN 978-81-8424-877-7

16. Chattopadhyay S, Karmakar S, Mitra M, Sengupta S (2012) Loss of phase fault detection of an induction motor. Int J Model, Meas Contr General Phy Electr Appl AMSE Series A 85 (2):18–34. ISSN 1259-5985

17. Chattopadhyay S, Karmakar S, Mitra M, Sengupta S (2012) Symmetrical components and current Concordia based assessment of single phasing of an induction motor by feature pattern extraction method and Radar analysis. Int J Electr Power Energy Syst 37(1):43–49. ISSN 0142–0615

18. Chattopadhyay S, Karmakar S, Mitra M, Sengupta S (2012) Assessment of crawling of an induction motor by stator current Concordia analysis. IET Electron Letter 48(14):841–842. doi:10.1049/el.2011.4008

# Index

## A
Acoustic analysis, 31
Air gap eccentricity, 48
Algorithm, 63, 91, 109, 122
Analytical tools, 154
Approximate coefficient, 126

## B
Bearing fault, 18
Bearings, 10
Blocked rotor, 25
Broken bar, 58
Broken rotor bar, 47, 58, 63, 91, 154
Broken rotor bar fault, 13

## C
Chemical analysis, 30
CMS, 155
CMS equations
    for total harmonic distortion factors, 45
CMS rule
    for highest order dominating harmonics
        determination, 45
CMS rule set, 42, 59, 89, 141, 146, 147, 149,
    151, 154, 156 *See also* for harmonic
    assessment in Clarke and Park plane, 45
    for unbalance assessment, 42, 44
Concordia, 59, 62, 63, 79, 89, 141, 148, 151,
    154
Concordia in Park plane, 88
Condition-based maintenance (CBM), 1
Condition monitoring, 25
Construction, 9
Continuous wavelet transform (CWT), 38, 114,
    117, 122, 123, 134, 154

Cooling fan, 11
Couple unbalance rotor, 16
CPU, 108
Crawling, 23, 147, 148, 154
CT, 139
Current analysis, 34
Current signature, 83
Current signature analysis (CSA), 4, 34, 35

## D
DAS, 108, 109, 143
Data, 35
Detail coefficient, 126
Discrete Fourier transform (DFT), 35
Discrete wavelet transform (DWT), 29, 39, 94,
    103, 114, 115, 117, 120, 154
During direct on line (DOL), 90, 108
Dynamic unbalance rotor, 17

## E
Electric Power Research Institute (EPRI), 57
Electrical stresses, 22
End flanges, 10
Envelope analysis, 66
Envelope detection, 68
Environmental stresses, 22
Existing techniques, 30

## F
Fast Fourier transform (FFT), 29, 30, 35, 36,
    79, 81, 83, 86, 110, 112, 117, 134, 154,
    156
Fault analysis, 30
Fault assessment, 35
Fault diagnosis, 1

Printed in the United States
By Bookmasters